Lecture Notes in Earth Sciences 61

Editors:
S. Bhattacharji, Brooklyn
G. M. Friedman, Brooklyn and Troy
H. J. Neugebauer, Bonn
A. Seilacher, Tuebingen and Yale

Springer

Berlin
Heidelberg
New York
Barcelona
Budapest
Hong Kong
London
Milan
Santa Clara
Singapore
Paris
Tokyo

Martin Breunig

Integration of Spatial Information for Geo-Information Systems

 Springer

Author

Dr. Martin Breunig
Institut für Informatik III, Universität Bonn
Römerstraße 164, D-53117 Bonn, Germany

G
70.2
B74
1996

Cataloging-in-Publication data applied for

Die Deutsche Bibliothek - CIP-Einheitsaufnahme

Breunig, Martin:
Integration of spatial information for geo-information systems
/ Martin Breunig. - Berlin ; Heidelberg ; New York ; Barcelona
; Budapest ; Hong Kong ; London ; Milan ; Paris ; Santa Clara
; Tokyo : Springer, 1996
  (Lecture notes in earth sciences ; 61)
  ISBN 3-540-60856-7
NE: GT

"For all Lecture Notes in Earth Sciences published till now please see final pages of
the book"

ISBN 3-540-60856-7 Springer-Verlag Berlin Heidelberg New York

© Springer-Verlag Berlin Heidelberg 1996
Printed in Great Britain

Typesetting: Camera ready by author
SPIN: 10528539   32/3142-543210 - Printed on acid-free paper

To Maximilian

Preface

The objective of this book is to introduce the practitioner as well as the more theoretically interested reader into the integration problem of spatial information for Geo-Information Systems. Former Geo-Information Systems are restricted to 2D space. They realize the integration of spatial information by a conversion of vector and raster representations. This, however, leads to conceptual difficulties because of the two totally different paradigms. Furthermore, the internal topology of the geo-objects is not considered.

In recent years the processing of 3D information has played a growing role in Geo-Information Systems. For example, planning processes for environmental protection or city planning are dependent on 3D data. The integration of spatial information will become even more important in the 3D context and with the development of a new generation of open GISs.

This book is intended to respond to some of these requirements. It presents a model for the integration of spatial information for 3D Geo-Information Systems (3D-GISs). As a precondition for the integration of spatial information, the integration of different spatial representations is emphasized. The model is based on a three-level notion of space that likewise includes the geometry, metrics and the topology of geo-objects. The so called *extended complex (e-complex)* is introduced as a kernel of the model. Its internal basic geometries are the point, the line, the triangle and the tetrahedron. It is shown how a *convex e-complex (ce-complex)* is generated by the construction of the convex hull and the "filling" of lines, triangles and tetrahedra, respectively. As we know from computer geometry, this results in substantially simpler geometric algorithms. Additionally, the algorithms gain by the explicit utilization of the topology of the ce-complex. This book also builds a bridge from the GIS to the object-oriented database technology, which will likely become a key technology for the development of a new generation of open Geo-Information Systems. In the so-called GEO-model kernel "building blocks" are introduced that simplify the development of software architectures for geo-applications. A geological application in the Lower Rhine Basin shows the practical use of the introduced geometric and topological representation for a 3D-GIS.

The initial ideas of this book were conceived when I was with the Institute of Computer Science at the Freie Universität Berlin during 1990 - 92. Most of the work, however, was done after my move from Berlin to Bonn at the Institute of Computer Science III (group of Armin B. Cremers, University of Bonn) in an excellent cooperation with the Geological Institute (group of Agemar Siehl) within a special research project[1]. The core of this book is my doctoral dissertation with the same title, written in German.

---

1. The "Sonderforschungsbereich 350" (SFB 350) is supported by the German Research Foundation (DFG).

First and foremost, I am extremely grateful to Armin B. Cremers for his help and the excellent support for the "GIS-group" at the Institute of Computer Science III. This book would not have been possible otherwise. In the same way I am indebted to Agemar Siehl from the Institute of Geology for agreeing to take over a second reader on short notice and for his support during my change to Bonn. His help in the translation of many geological terms is also gratefully acknowledged. I explicitly thank the speaker of the special research project, Horst Neugebauer, its board and members for the warm-hearted reception. Horst Neugebauer also stimulated me in writing this English manuscript. I thank Lutz Plümer for his useful comments on the manuscript and Ralf Hartmut Güting for his stimulating critics. My thanks also go to Hans-Jörg Schek (now ETH Zürich) who sparked my interests in Geo-Information Systems during the DASDBS-Geokernel project at the Technical University of Darmstadt.

I am indebted to my colleagues Walter Waterfeld at the Technical University of Darmstadt, Axel Perkhoff, Gernot Heyer and Peter Schütze at the Freie Universität Berlin and my colleagues at the University of Bonn for the always pleasant atmosphere. I particularly thank Thomas Bode for his extensive involvement in the SFB and for many discussions about GIS and database problems. He and Andreas Bergmann, Jürgen Kalinski, Thomas Kolbe, Gerhard Lakemeyer and Wolfgang Reddig have made their comments on parts of this book. I also thank Rainer Alms and Christian Klesper for their patient explanation of many geological facts.

My thanks go as well to Ralf Noack, Ingo Schoenenborn, Jürgen Stienecke and Norbert Klein, who had their part in the successful outcome of this book with their M.Sc. theses and other work. Finally I am indebted to my parents to enable me the study of computer science.

Bonn, November 1995   Martin Breunig

Contents

Chapter 1

Introduction

1.1 Geo-Information Systems

Today Geo-Information Systems (GISs) are applied more and more often in agencies, universities and in industry. Their goal is to support spatial planning processes such as the city management or the environmental management. Hereby the processing of 3D information plays an increasing role (RAPER 1989). Today's GISs are pure 2D systems. They are not prepared for the management and processing of three-dimensional data[1]. Furthermore, information from different data sources like geoscientific maps, soil information systems or climate tables are often used for planning processes. Last, but not least, their integration runs into difficulties because of their different representations[2]. Let us first introduce into the field of Geo-Information Systems.

1.1.1 Historical Development

The predecessors of today's GISs aimed at the automation of special manual tasks like the drawing of maps. For their support, tools were developed for the efficient management of spatial data like spatial access methods (TAMMINEN 1982; GUTTMAN 1984; NIEVER-GELT; HINTERBERGER 1984 etc.) and spatial representations (SAMET 1990a, 1990b) since the beginning of the 80ties. Parallel to that, first commercial products for the support of map construction like ARC/INFO (MOREHOUSE 1985), TIGRIS (HERRING 1987), SPANS (KOLLARITS 1990), SYSTEM 9 (1992), THEMAK2 (GRUGELKE 1986) etc. were developed, which were designated as "Geographical Information Systems" (GISs)[3]. However, it became clear, that the functionality of these systems was strongly restricted and cutted to special applications. Since the middle of the 80ties the database community tried to deal with the demands of so called non-standard applications like GISs. The result was the development of extensible database systems like GENESIS (BATORY et al. 1986), DAS-DBS (SCHEK and WATERFELD 1986, PAUL et. al. 1987, WOLF 1989, WATERFELD and BREUNIG 1992), PROBE (DAYAL et al. 1987), PRIMA (HÄRDER et al. 1987), EXODUS (CAREY et al. 1988), AIM-P (LINNEMANN et al. 1988), GRAL (GÜTING 1989), STARBURST (HAAS and CODY 1991) POSTGRES (STONEBRAKER and KEMNITZ 1991) and OMS (BODE et al. 1992). GÜTING (1988) made a contribution to the extension

---

1. The "3D-GISs" being on the market essentially are CAD systems which do not allow the connection of thematic and geometric data. This, however, is an essential point for GISs as we will see in the following.
2. Vector and raster, respectively.
3. For a comparison of the systems see also (BÜSCHER et al. 1992).

of relational systems with his "georelational algebra" and the Gral system (GÜTING 1989). In this work the relational algebra is extended by geometric 2D operations. Historically seen in a "bottom-up" way, first proposals for query languages and user interfaces for GISs followed not before the end of the 80ties and at the beginning of the 90ties (NEUMANN 1987; EGENHOFER and FRANK 1988; VOISARD 1991; VOISARD 1992; SVENSSON et al. 1991; BOURSIER et al. 1992). Furthermore, extensible and object oriented prototype systems like GEO^{++} (VAN OOSTEROM and VIJBRIEF 1991; VIJBRIEF and OOSTEROM 1992); SMALLWORLD GIS (1993), GeO$_2$ (DAVID et al. 1993), GODOT (GAEDE and RIEKERT 1994; EBBINGHAUS et al. 1994) were developed. The extensiblity of the GIS functionality was also taken into consideration in the GIS GRASS (1993) and in the "object oriented" SMALLWORLD GIS (1993). The ROSE-Algebra (GÜTING and SCHNEIDER 1993) provided the embedding of geometric 2D operations into object oriented query languages. GÜNTHER and LAMBERTS (1992) give an insight into the use of object oriented databases for geographical and environmental applications. Theoretical work mainly deals with topological and geometric data models in 2D space (EGENHOFER 1989a, 1989b, 1991; PIGOT 1991, 1992, 1994; WORBOYS and BORFAKOS 1993). Recently also the modeling and management of 3D and 4D data is demanded (WORBOYS 1992; DE HOOP et al. 1993; FRANK 1994; HEALEY and WAUGH 1994; PILOUK et al. 1994; VAN OOSTEROM et al. 1994; WORBOYS 1994). Besides additional functionality for the analysis of spatial data (GOODCHILD 1990; RAPER and RHIND 1990; BURROUGH 1990; CHOU and DING 1992; RAPER 1992; VAN OOSTEROM et al. 1994), models and tools for the interoperable use of GISs and for the *integration of spatial information* are demanded (GUENTHER and BUCHMANN 1990; WORBOYS and DEEN 1991; SALGE et al. 1992; SCHEK and WOLF 1993; WOLF et al. 1994). The motivation is the increasing number of networks and the resulting growing of data exchange. The integration plays an even more important role, if spatial information from different systems with potentially different hardware platforms are to be managed and processed.

1.1.2 Definitions

The acronym "GIS" used in the sense of a Geographical Information System was introduced by R. F. TOMLINSON (1972) at a compendium of the International Geographical Union Commission on Geographical Data Processing and Sensing in Canada. Tomlinson designates a GIS as *"not a field in itself but rather the common ground between information processing and the many fields utilizing spatial analysis techniques"*. CLARKE (1986) gives a definition that has its validity until today. He designates a GIS as *"computer-assisted systems for the capture, storage, retrieval, analysis, and display of spatial data"*. Thus the focus is set on the management, the processing and the output of spatial data.

COWEN (1988) gives an overview of different GIS definitions. He devides them into the four approaches *"process-orientated approach"*, *"application approach"*, *"toolbox approach"* and *"database approach"*. The process-oriented definition sees a GIS as an information system that consists of subsystems which convert geographical data into "useful information" (TOMLINSON 1972; CALKINS and TOMLINSON 1977). The application definition paraphrases a GIS within its respective application field. A GIS is seen as a system e.g. for city planning etc. (PAVLIDIS 1982). The toolbox definition (DANGERMOND 1983) emphasizes that a GIS consists of complicated procedures that are hidden from the user. With the procedures, algorithms for the processing of spatial data are executed. Finally the database oriented definition describes a GIS as a toolbox that communicates with a database (GOODCHILD 1985). With this definition a GIS is a system that uses a spatial database for answering geographical queries. MAGUIRE et al. (1991) and RHIND (1992) also use the acronym "GIS" like the already mentioned authors in the sense of a "Geographical Information System". Today the term "Geographical Information System" is mainly used to distinguish pure two-dimensional GISs from three-dimensional GISs (TURNER 1992). To emphasize the extension of its tasks in the third dimension, for instance in geology or geomorphology, the term "Geoscientific Information System" GSIS) was coined (VINKEN 1988). Later this term was revised into the shortform "Geo-Information System" (GIS) (VINKEN 1992, KELK 1992). By this an information system for the different geoscientific disciplines was meant (compare VINKEN 1992). Since then in the German usage the term "Geo-Informationssystem" was mostly used (BILL and FRITSCH 1991; SIEHL 1993). Alternatively the term "Raumbezogenes Informationssystem" (RIS) is in use (BARTELME 1989; GÖPFERT 1991) and in the english language sphere the term "Spatial Information System" (SIS) (PIGOT 1991; ABEL et al. 1992; LAURINI and THOMPSON 1992) is used. PIGOT (1994) uses the term "Spatial Information System" as a collective name for "Geographic Information Systems" (GISs), "Computer-Aided Design Systems" (CAD) and visualization systems for full three-dimensional applications. We designate a GIS in the sense of a "Geo-Information System". Geo-Information Systems often are subdivided into several classes according to their application fields like Geographical Information Systems (GISs), Environmental Information Systems (EIS) or Land Information Systems (LIS).

Definition 1.1: By a *Geo-Information System (GIS)* we understand a system for the in-/output, management, processing and *integration* of geoscientific, space refered[1] information. The GIS can be extended by the fourth dimension (time). Thus a Geo-Information System is a space refered information system for the geosciences.

1.1.3 Types of GISs

GISs often are subdivided into different categories according to the type of the representation in which their spatial data are managed. Hitherto the subdivision is restricted to the representations in 2D space. At the end of this chapter, however, we also refer to 3D-GISs.

Vector-based GIS:

Most of today's GISs like ARC/INFO (SCHALLER 1988), SYSTEM 9 (1992), GEOVIEW (SINHA and WAUGH 1988), GEO++(VAN OOSTEROM and VIJBRIEF 1991; VIJBRIEF and OOSTEROM 1992) etc. mainly use vector data, because they are easy to be digitalized and can be processed in the GIS as layers or objects. Additionally, in these systems raster data, for instance from photogrammetry, or pictures can be processed by "underlying" the vector maps in the GIS. An example for that is the combination of a digitalized vectorial overall view plan with a complete topographical map.

---

1. The term "space refered" includes pure thematic information and spatial 3D-information.

Raster-based GIS:

Refering ZHOU and GARNER (1991), a raster- or image-based GIS (IGIS) is characterized by the following peculiarities:

- it supports the geographical data handling and manipulation (thematic attributes),
- it is raster-based, i.e. raster algorithms are supported,
- it has efficient data interfaces for the conversion of vector- and raster-structures,
- the results from the GIS can again be converted into the vector format.

The commercially available raster-based GISs like SPANS (KOLLARITS 1990), GRASS (1993), SMALLWORLD GIS (1993) etc. are originated from image processing systems. Refering ZHOU and GARNER (1991), a RIGIS is an IGIS that uses a relational DBMS for the management of the attribute data.

Hybrid GIS:

Because of the heterogeneity of the data in industry, authorities and scientific establishments, today such GISs are particularly required that base upon multiple representations. Most of the systems we have enumerated as vector-based GISs like ARC/INFO are hybrid GISs. The different spatial representations are used "simultaneously" in hybrid GISs, i.e. a map overlay, for instance, of vector and raster data is possible. ZHOU and GARNER (1991) have realized a RIGIS that uses ARC for the management of the vector data, INFO for the thematic data and ReMAP for the raster data.

To enable descriptive spatial queries, independently of the internal spatial representation on the database in hybrid GISs, SQL (structured query language) can be extended by spatial predicates. Approaches of this kind became well-known by the work of GÜTING (1988) with his "Georelational Algebra", of HERRING et al. (1988) with extensions of an object oriented GIS, of ABEL (1988) with SQL-based spatial extensions of a relational database (SIRO-project), of INGRAM and PHILLIPS (1988) with spatial extensions of a hybrid GIS-data model and of EGENHOFER (1989) with XSQL, an SQL extension for spatial relationships. For interactive queries in GISs, however, such SQL-extensions of relational database management systems have proved to be inefficient because of the relational, i.e. "flat" representation of the geometric data and the lacking of spatial access support.

In connexion with hybrid GISs the question arises for the availability of data in suitable formats. RHIND and GREEN (1988) made a study with regard to the interpretation of GIS functionalities depending on different spatial representations. A result of the study was that one of the representations which is mostly used in GISs, namely the so called *"topological model"*[1] is fewest available in today's data sources. BILL and FRITSCH (1991) remark that the data of the different components of an information system (hardware: 3 - 5 years, software: 7 - 15 years, data: 25 - 70 years!) have the longest "surviving time". Thus the data often are older than the "data models[2]" being developed in the last 20 years. The conformity to "new"

---

1. I.e. a vector representation that considers the topology of areas.
2. In the geoscientific GIS literature spatial representations like vector and raster are often called "data models".

"data models", i.e. spatial representations, is an extremely time-consuming process because of the uge sets of collected data and thus it is often tackled in delay.

3D-GIS:

A closer look to the data handling of today's commercial GISs shows that the third dimension of space is at most "carried along" as a thematic attribute, i.e. for example as a latitude value in isoline maps. It is confusing that in geography "3D" is often interpreted as "2D + time", i.e. the management of two-dimensional geometries at different times. Some vendors of GISs also call the carrying along of an arbitrary thematic attribute of a two-dimensional geometry as "3D"-representation. Furthermore, it is unclear which spatial representations will put through for future 3D-GISs. Hitherto representations like "CSG" or the boundary representation, which are well known from CAD, are used for the visualization of 3D-objects. However, in these systems, a management of geometric objects is missing, because mostly the graphical primitives only consist of single edges.

First theoretical approaches for 3D-GISs particularly deal with topological data models (PI-GOT 1991, 1992, 1994; DE HOOP et al. 1993; PILOUK et al. 1994). Numerous resaerch work tried to solve the problem of modeling and managing 3D-data (VINKEN 1988, 1992; TURNER 1992; RHIND 1992; KELK 1992; ABEL and OOI 1993; BRUZZONE et al. 1993; DE FLORIANI et al. 1994; HACK and SIDES 1994; HEALEY and WAUGH 1994; PI-LOUK et al. 1994; VAN OOSTEROM 1994). However, hitherto no drastic success could be achieved. The development of 3D-Geo-Information Systems still stands at the very beginning. All the same two directions in the development become already apparent: The first tries - starting from the visualization of three-dimensional geometries and originally coming from the CAD field - to stamp an efficient, interactive modeling and visualization component on the 3D-GIS (MALLET 1992; RAPER 1992; TURNER 1992; PFLUG et al. 1993; SIEHL 1993). Against this, the second direction has the primary goal to develop a 3D-data management and to couple it with 3D-modeling and visualization tools (SCHEK and WOLF 1993; BODE et al. 1994; VAN OOSTEROM et al. 1994; WOLF et al. 1994). This direction basically originates from the database side. A closer coupling of the two directions could lead to a new architecture for a 3D-GIS.

One of the most important applications for 3D-GISs is geology. Geology, for its part, renders its own service for the development of 3D-GISs, because it formulates new requirements, formalizes geological processes (BURNS 1975; SAKOMOTO 1994) and interacitvely models geological areas and volumes (SIEHL 1988; PFLUG et al. 1992; SIEHL 1993).

1.1.4 Architecture of GISs

The architecture of today's GISs can be subdivided into three levels (see fig. 1.1). The *GIS-user interface* mostly consists of a macro language, of an editor for the data input and a graphical user interface with menus. The *GIS-tools* provide some geometric operations like distance, buffer generation around geometric objects or overlay of maps. Furthermore, advanced analysis operations for certain application fields are provided.

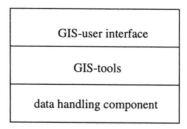

Fig. 1.1 Architecture of a GIS

Depending on how the *data handling component* is structured, VIJLBRIEF and OOSTE-
ROM (1992) distinguish three types of architectures (see fig. 1.2a-c), namely a *dual architecture,* a *layer architecture* and an *integrated architecture.* We add a fourth type of architecture, the *object oriented architecture* (see fig. 1.2d).

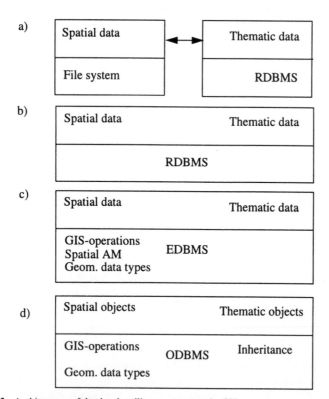

Fig. 1.2   Architectures of the data handling component of a GIS
           a) Dual architecture,
           b) layer architecture,
           c) integrated architecture,
           d) object oriented architecture.

Dual architecture:

With the dual architecture, which was also called "combination architecture" by WATER-FELD (1991), thematic data are managed by an RDBMS and spatial data are directly handled by the file system. The relationship between the data being managed by the two different systems is realized by pointers from the thematic data to their geometries. Examples for GISs with a dual architecture are ARC/INFO (MOREHOUSE 1985) und SICAD (SONNE 1988). This variant of architecture has advantages in efficiency. However, because of the separated storing of thematic and spatial data, more accesses on the data are necessary. The user has to compose his/her geo-objects from the thematic and spatial data by his/her own, i.e. the pointers between thematic and spatial data are visible for the user.

Layer architecture:

With the level architecture, or "additional level architecture", as it is called by WATER-FELD (1991), thematic and spatial data are managed by an additional level on top of an RDBMS. An example of a system with a layer architecture is SYSTEM 9 (BURGERMEI-STER 1990). This variant of an architecture has the advantage that the transaction management of the RDBMS can be used for all data, i.e. as well for thematic data as for spatial data. The use of the data is transparently, equal if the data are thematic or spatial. A disadvantage is the lacking efficiency, as very large geometries can only be managed if they are "cut into pieces". Furthermore, spatial access methods have to be realized in the user interface of the RDBMS.

Integrated architecture:

With the integrated architecture, an extensible DBMS (EDBMS) takes over the efficient management of thematic and spatial data. The user can extend the EDBMS by geometric data types, spatial access methods or simple GIS operations that directly support GIS-tools. One of the first systems with an integrated architecture was TIGRIS (HERRING 1987). A further representative is GEO^{++} (VAN OOSTEROM and VIJBRIEF 1991) that is put on top of POSTGRES (STONEBRAKER and KEMNITZ 1991). The THEMAK2 system (GRUGEL-KE 1986) that was put on the DASDBS Geokernel can also be seen as a component of a GIS with an integrated architecture (WATERFELD and BREUNIG 1990). The main advantage of an integrated architecture is the good performance that is achieved by the embedding of user defined geometric data types and spatial access methods at al low system level.

Object oriented architecture:

The object oriented architecture uses an object oriented database management system (ODB-MS) for the common management of thematic and spatial objects. As the programming language interface of the ODBMS provides data encapsulation incl. inheritance and polymorphism, geometric data types with special GIS-operations can be directly defined in the ODB-MS. However, in today's commercial ODBMSs, spatial access methods can only be realized "on top" of the ODBMS, because there are no corresponding interfaces in the lower system levels. Only one-dimensional indices of the ODBMS can be exploited, for example to realize quadtrees (SAMET 1990a) "on top" of the ODBMS (see GAEDE and RIEKERT 1994). GO-DOT (GAEDE and RIEKERT 1994) and GeO$_2$ (DAVID et al. 1993) are examples for GISs with an object oriented architecture. GEOSTORE (BODE et al. 1994) can be seen as a component of a GIS with an object oriented architecture. The object oriented architecture basically distinguishes itself from the "layer based" systems in the kind of the data management. Instead of thematic layers of maps, in the object oriented architecture single objects with specific GIS-operations can be managed and processed.

1.2 Integration of Spatial Information

1.2.1 Motivation

The support of new, so called non-standard database applications, for instance in the engineering or in the geosciences, is a central theme of the database research. Additionally, many of these non-standard applications demand the integration of their data and models (ANDERL and SCHILLI 1988; FONG and GOLDFINE 1989; SILBERSCHATZ et al. 1990; PIWOWAR et al. 1990; ACM 1990; KREMERS 1991; BRESNAHAN et al. 1992; ABEL and OOI 1993; BODE et al. 1993; SCHEK and WOLF 1993 etc.). In recent time the integration of heterogeneous data and the cooperation of different systems with the goal of solving single tasks has developed to a new field in the database research. It is supposed to lead to so called *interoperable systems*. In many of the database applications such as CAD, civil engineering, mining, petroleum exploration, environmental protection, medicine etc., spatial information play a central role. There is a growing need for the *integration of spatial information* in a central logical data model, for instance for a further processing in CAD- and planning software systems (MEISSNER and WÖRNER 1992) or in geoscientific applications (FGDC9 1994). Thereby the explicit representation of *topology* increases in importance. In geo- or environmental-information systems, for example, digitalized vector data shall be integrated with raster data from photogrammetry to execute environmental analysis. However, the process of integrating spatial information causes problems because of the different spatial representations (raster/vector in 2D space). This is even more true for 3D-applications.

SHEPHERD (1991) gives an overall view of the role of GISs as "information integrator". The demand for the integration of data and models will still grow in future because of the development of multimedia GISs. The interdisciplinary integration of spatial information also plays an important role for the international cooperation, for instance for the management of environmental problems in the European Community (SALGE et al. 1992).

1.2.2 Requirements for a 3D-GIS

Today, planning processes like the city planning or environmental monitoring, require 3D-information in an increasing extent. With the management of the third dimension, however, we predominantly enter new ground in the database/GIS field. Previous GISs like ARC/INFO, SPANS, TIGRIS, SMALLWORLD GIS etc. mainly provide only tools for the processing of 2D geometries.

We now informally introduce some terms that will follow us in the following. According to NEUMANN (1987) a *geo-object* represents a - either in "reality" or in an imagining world - existing geoscientific object, such as a city or a river. A geo-object may be composed by a name, a geometric description and a thematic description (see fig. 1.1). We count all geometric and topological attributes and methods to the geometric description. Under consideration of the topology we call the geometric description also *spatial description*. The spatial description (see also MATSUYAMA et al. 1984) provides information about the shape and the position of a geo-object. All its non-spatial attributes and methods, such as its database key and the non-spatial properties like the name, the context-specific composition of a geo-object etc., belong to the *thematic,* i.e. non-spatial *description*. We define that a geo-object must include *at least one geometric attribute or one geometric method*. Geo-objects with common properties can be generalized in a *geo-object class*.

By the topology[1] of a single geo-object we informally understand its internal topological structure. We also speak of the *local topology* of a geo-object. For instance, if the geometry consists of an area partition, the topology describes the neighbourhood between the single faces. By the *topology of several geo-objects* or the *global topology*, we mean the relative spatial position of the geo-objects between each other, like intersection or adjacency. As we will see in chapter 4, the topology between geo-objects can be specified with the topological concepts *"boundary"* and *"interior"*. Let the boundary of a node be the node itself, the boundary of an edge its start- and end-node, the boundary of an area its circumscribing polygon and the boundary of a volume its surface, respectively. Informally the *geometry of a geo-object* is its spatial extension given by its coordinates. The geometry can be of different dimension and it can be represented by different *spatial representations*.

By a *spatial information* (see fig. 1.3) we understand an information being gained from the spatial description of a geo-object, such as the position of a geo-object in space. Against this an *information with spatial context* includes the information which is gained by the thematic description of the geo-object. A spatial information can for instance be gained by the following query: "Provide all geo-objects that intersect the geo-object *A* ". An information with spatial context we gain for instance with the following query: "Provide all cities within North Rhine Westphalia that have more than 100000 inhabitants and that have a distance of not more than 100km from Bonn".

---

1. The term "topology" is formally defined in chapter 4.

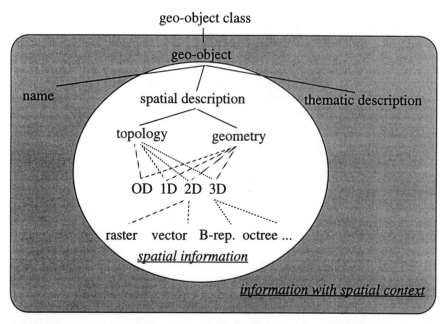

Fig. 1.3 Geo-object, spatial information and information with spatial context

Obviously the following hitherto unsolved problems stand in the way of a unified use of spatial information in a 3D-GIS: On the one hand the spatial data models are not yet suitable for 3D-GISs. On the other hand there is still no solution for a unified use of heterogeneous spatial 3D-representations. The integration of spatial information leads to the following essential requirements for the spatial data management and processing of a 3D-GIS:

(1) The management and processing of *complex geo-objects,*
(2) an efficient spatial access to geo-objects with *heterogeneous spatial representations,*
(3) the efficient support of *spatial operations* executed on the geo-objects.

Requirement (1) claims the management of complex structured objects, as today's ODBMSs provide them, but only few object oriented GISs like GEO[++] (VIJLBRIEF and OOSTEROM 1992) or GeO$_2$ (DAVID et al. 1993) . In today's GISs, requirement (2) is only provided for special spatial operations. Examples for requirement (3) are the area search (window query), distance operations between points and special topological relationships like the neighbourhood in area partitions. In raster-oriented systems like SPANS (KOLLARITS 1990), additional operations of the "cartographic algebra" can be found like the addition and subtraction of layers or filter operations. Today's available GISs mostly are closed systems that only provide file interfaces to foreign systems. Furthermore, the data exchange is made more difficult due to the heterogeneous formats of the different GIS vendors.

1.3 Goals and Overview

We will introduce an integration approach that considers as well the *third dimension* of space as the *geometry and topology* of heterogeneously represented spatial information. The integration of spatial information is treated by a general model that contains the specialized spatial representations. This point is particularly interesting with regard to future *inter-operable Geo-Information Systems*. From the requirements for 3D-GISs the following central questions of the field result:

(1) Representation of the 3D-geometry and topology of geo-objects,
(2) Integration of spatial representations,
(3) integrated management and processing of geo-objects with heterogeneous spatial representations.

The book is organized as follows: After some introducing considerations about Geo-Information Systems and the motivation for the integration of spatial information in chapter 2, spatial representations in 2D and 3D space are examined according to their suitability for the use in a 3D-GIS. Hereby the problem of conversion is emphasized. Chapter 3 introduces approaches for the abstraction of spatial representations and tries to value them according to the utility with an integration of spatial information. In chapter 4 a model for the integration of spatial information is introduced that uses the so called e-complex, an extended simplicial complex, as spatial representation. Efficient geometric algorithms and an algebra of spatial operations on e-complexes are introduced. In chapter 5 the embedding of the e-complexes in an object oriented GEO-model kernel for a 3D-GIS is discussed. In chapter 6 measurements with e-complexes shall underpin the efficiency of a geometry and topology representation for 3D-GISs that is based on e-complexes. In chapter 7 a geoscientific application of the Lower Rhine Basin is introduced. Chapter 8 summarizes the results and gives an outlook to ongoing work in the field.

Chapter 2

Spatial Representations

In this chapter we intend to have a closer look to spatial representations that may be a candidate for a unified representation of spatial data in a 3D-GIS. Today, as the transition from 2D- to 3D-GIS is attacked, we can base on existing 3D representations of computer graphics. This is similiar to the beginning of the GIS history, when one could fall back upon the 2D representations used in that time. However, the different requirements, namely visualization vs. data management, must also be weighed in our judgement. For the comparison of the advantages and disadvantages of the spatial representations we mainly orientate ourselves at the relevant literature (REQUICHA 1980; REQUICHA and VOELKER 1982; MEIER 1986; RHIND et al. 1988; MÄNTYLA 1988; HOFFMANN 1989; BILL and FRITSCH 1991; MAGUIRE et al. 1991).

2.1 Representations in 2D Space

Today's GISs like ARC/INFO, SICAD, SPANS, TIGRIS, SMALLWORLD GIS, THE-MAK2, GRASS, SYSTEM 9 or GEO^{++}, support - though still very limited - as well the *vector* as the *raster representation*. Raster representations serve as background pictures for digitalized maps. In this context we speak of so called *hybrid 2D-representations* (see chapter 1).

2.1.1 Vector Representation

The vector representation is often designated as "vector data model" (PEUQUET 1984) or "vector format" (WORBOYS and DEEN, 1991). In connection with geo-objects WORBOYS and DEEN (1991) speak of an "object-based spatial model". Important is that the geometry may be attached to objects.

- Geometry and topology in the vector representation

In the vector representation points are the presentative of the geometric information. All further geometries (lines, polygons etc.) can be composed of points. Geometric attributes of geo-objects like the length, area distance etc. can be derived from the coordinates of the points. However, at the same position in space two different points can exist, i.e. points in the vector model are not unambigious. The topological relationship "to be connected" connects a start point with an end point of an edge. Areas are defined by their boundary polygons (bounding edges). They are represented as a list of points or a list of edges, respectively.

- Different realizations of the vector representation for area networks

In the field of GISs often vectors are used to represent topologies in area networks over the neighbourhoods of the edges. Important application fields are for example land registers or the reparcelling of the fields in a village (see BILL and FRITSCH 1991).

According to the way the vectors are stored, the following area-network representations can be distinguished (see RHIND and GREEN, 1988):

- non-topological representations (e.g. the "Spaghetti-model": see BRASSEL 1983, PEUQUET 1984),
- simple topological representations,
- directed topological representations (e.g. GBF/DIME; U.S.CENSUS 1970),
- hierarchically indexed topological relationships (e.g. POLYVRT: PEUCKER and CHRISMAN 1975; TIGER: MARX 1984; ATKIS: AdV 1989).

The order of the representations is to be seen as a historical development. Thus the next representation is always an improvement of its predecessor. In the "spaghetti model[1]" the geometries are modeled without consideration of spatial neighbourhoods. Especially the boundaries of adjacent polygons are stored twice. Against that, in the topological model elementary spatial relationships like neighbourhoods are included. Thus points, lines and polygons can be stored without redundancy. If directed edges are introduced like in the Geographic Base File/Dual Independent Map Encoding, an unambigious assignment of a left and a right area for each edge is provided. This is especially advantegeous for adjacent polygons. But in the GBF/DIME-representation no reference from the polygon to its edges is provided. Neighbourhood relationships are also considered in the indexed vectors like the POLYgon-ConVeRTer. In this representation hierarchical polygons are composed of lines and lines are composed of points. A modification of this representation is used in ARC/INFO. The same principle is utilized in the TIGER-representation which was developed by the U.S. Bureau of the Census. TIGER introduces additional points for the navigation in the structures such as the access on the next edge or the next "mesh", i.e. the next polygon. The ATKIS representation (AdV 1989) additionally provides a hierarchical modeling of objects and it considers a raster-based modeling as well. An overview of different vector representations is also given in FINDEISEN (1990).

- Advantages and disadvantages of the vector representation

The advantages of the vector representation are the high accuracy, the low need for storage place, the simple execution of coordinate transformations and the simple realization of distance operations like the Euclidean distance.

---

1. In the geoscientific literature spatial representations like raster or vector are often designated as a "data model" (PEUQUET 1984) or "data structures" (PEUQUET 1984; RHIND et al. 1988). However, in computer science, spatial access methods like the GRID FILE (NIEVERGELT and HINTERBERGER 1984) or the R-tree (GUTTMAN 1984) are also called "spatial data structures" (WIDMAYER 1991).

As a disadvantage the difficult computation of intersections and neighbourhoods in the general case were showed off (MEIER 1986)[1].

2.1.2 Tesselating Representations

Tesselating representations are decomposing the plain into discrete cells. They are also designated as tesselation models or "layer-based models" (WORBOYS and DEEN 1991). The proceeding is exactly vice versa to that described for the vector representation: not geometries are generated from points and attached to objects, but the plain is decomposed into independent cells. Whereas the vector representation starts from "spatial objects", the tesselating representations describe the spatial continuum. However, a posteriori, an assignment from objects to geometries is also possible for tesselating representations, for example by means of a classification of raster cells. Tesselating representations can be classified in the following way (compare RHIND and GREEN 1988):

- regular tesselations (e.g. grid or raster),
  - nested regular tesselations (e.g. quadtrees),

- irregular tesselations (e.g. triangulated irregular networks TINs and Voronoi diagrams, respectively),
  - nested irregular tesselations (e.g. point quadtrees, k-d trees).

At the regular tesselations all partitionings are equally and they are consisting of regular polygons. The best known tesselations are the square, the triangle and the six-gon. Whereas regular tesselations are "absolutely" determined a priori, irregular tesselations are "dependent on the data", i.e. the tesselation is generated depending on the distribution and the order of the data.

- <u>Raster representation</u>

The raster representation traditionally is the most used tesselating representation in the geosciences. It is also designated as "raster data model" (PEUQUET 1984). Fig. 2.1 shows the boundary of a Berlin map in the vector and the raster representation.

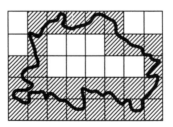

Fig. 2.1 Berlin map (sketch) in the vector and the raster representation[2]

---

1. Usually $O(n^2)$ according to the number n of the participating lines.

The term *raster* is often used for two different facts. On the one hand, with a raster a grid is meant, which interpolates measure points on an area. This raster we call *geo-raster* in the following. To each cell of the geo-raster a set of thematic attributes is assigned, like "ground exploitation", "precipitations" etc. With a classification[1] of single raster cells, geometric objects can be generated from single cells of the geo-raster. On the other hand, the raster used in computer graphics is well known. In this *"graphic-raster"* the number of raster cells (pixel) determines the graphical solution of the image. A cell consists of a set of colour tables (e.g. of a satellite image) for the graphical representation. If we decreased the size of the cells of the geo-raster against zero and if we additionally replaced the thematic attributes by colour tables, then the geo-raster (grid) would become a graphic-raster.

- Geometry und topology in the raster representation

The raster cell is the only "object type" in the raster representation, there is no explicit geometry of an "object". For the generation of geometries the properties (attributes) of an "object" are abstracted. Of course the "objects" have to be generated in advance. The attributes are seen as homogeneous within the rectangularly and regularly ordered cells. Thus they can be composed as thematical equal cells. Therefore all the cells belonging to one object have to be uniformly marked. The only geometric basis element of the raster representation is the *raster cell*, which is rectangular, has an area-like structure and covers an area with a homogeneous thematic meaning. All raster cells have the same size and are regularly arranged. Lines are represented as connected sequences of raster cells and areas are represented as sets of raster cells.

- Advantages and disadvantages of the raster representation

The advantages of the raster representation are simultaneously the disadvantages of the vector representation and vice versa. Most of the topological properties are implicitly given, i.e. the order-, neighbourhood- or adjacency-properties can be directly determined from the cells and their neighbours. The sum and the difference of elementary areas can be easily computed, as raster operations are extremely simple. The spatial solution, however, has to fit itself in the smallest representable element. The large need for storage place can be reduced by compression techniques. Rotations and coordinate transformations are difficult to compute. Note that for the raster representation many such operations are necessary, namely one operation for each raster cell . A further disadvantage of the raster representation is that "objects" must be generated by costly methods (e.g. component labeling) (DIKAU 1992).

2.1.3 Hybrid Representations

Hybrid spatial representations[2] aim at a raster strategy for the gross decomposition, but in detail the vector representation is used. Thus geometric operations on objects can be execu-

---

2. The raster was intentionally chosen very gross to emphasize its principle and to show its difference with the vector representation.

    1. For example with the well known "component labelling" method that marks the raster cells, which correspond to a geo-object.
    2. In the geoscientific literature also called "hybrid models".

ted in a raster cell. If we manage the object geometries themselves as vectors in the leaves of a quadtree (SAMET 1990a, 1990b) and not only their quadtree approximation, then the quadtree can be seen as a hybrid representation (see fig. 2.2).

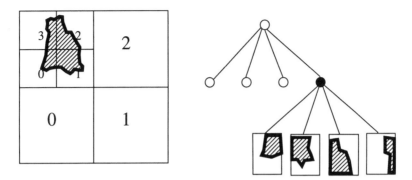

Fig. 2.2 Geometry of the former West-Berlin as quadtree

In the DIGMAP-System (DANN and SCHULTE-ONTROPP 1989), for example, that can be seen as a component of a GIS, drillings are represented in the vector representation and mines are represented as parameters. Complex geometric objects like stratigraphic and geological areas can be represented as rasters.

With the hybrid representation, a compromise is made, that reduces the disadvantages as well as the advantages of the vector and the raster representation.

2.1.4 Analytical Representations

To complete the list, we also mention the analytical representations, which represent geometries by means of function equations (e.g. splines) or by parameters. These representations do not play a significant role in general, as in geo-applications we often have very complex and irregular objects that can only by insufficiently modeled with these techniques, which originate from mechanical engineering.

The main advantage of the analytical representations is the accurate representation of curved geometries. The geometric algorithms (e.g. boundary integrals), however, are very costly. In table 2.1 the raster, vector and analytical representation are compared.

Spatial Repr.\ Real World Geom.	Geo-Raster	Vector	Analytical
	Accurate	Accurate	Accurate b a
	Approx.	Accurate	Accurate h a
	Approx.	Approx.	Accurate M r
	Approx.	Approx.	Accurate M r1 r2

Table 2.1 Simple area-like geometries in 2D space represented in the geo-raster, vector- and analytical representation

In many geoscientific applications good approximations for irregular geometries are needed. If we consider table 2.1 column by column, we see that in the geo-raster representation only such geometries can be accurately represented that are spatially oriented with the x- and y-axes (as far the spatial solution allows it). Triangles, for example, must already be approximated. Against this, in the vector representation an approximation must only be provided for geometries with curved elements. Whereas with the analytical representation all geometries can be accurately represented, in the vector representation "holes" and "islands" can be only approximately represented. Furthermore, additional information has to be provided, if the parts of the geometry are "inside" or "outside" the geo-object. With the ROSE-algebra (GÜTING and SCHNEIDER 1993), for example, polygons with holes can be represented. The raster representation enables such a representation only by means of an approximation. With the analytical representation, several parameters are necessary to solve the problem. Furthermore, the information has to be added, where "outside" and "inside" of the geometry is, as we indicated with the ring in the right corner at the bottom of table 2.1.

2.2 Representations in 3D Space

In CAD "3D-models" are grossly divided into wireframe-, surface- and solid models (RE-QUICHA and VOELCKER 1982). This subdivision describes "what You see at the screen" using these models. A cube, for example, in the three models is represented by its edges (wireframe), its faces (surface) and its solid (volume). Our goal, however, on the search for a spatial representation being adequately for the data management of a 3D-GIS, is to see the spatial representation independently of the graphical representation[1]. Analogous to the 2D space we subdivide the 3D representations into vector-based, tesselation-based, analytical and hybrid representations.

2.2.1 Vector-based Representations

- Wireframe representation

As a first representation we name the wireframe representation, which represents the geometries of objects by line segments or by curves. However, in this representation we cannot attach information to surfaces and solids. Thus the wireframe representation is unusable for the computation of intersections between solids and therefore is out of question as a representation for a 3D-GIS.

- Boundary Representation (B-Rep)

In the boundary representation the geometries of the objects are represented by their boundary elements. They may be polygons, edges, nodes, but as well analytical functions and areas that are defined by interpolations or approximations (e.g. Bezier- or B-spline-areas). We first consider the first case[2] and call the corresponding representation the "polygonal vector representation" (PVR). The wireframe representation is a special case of the boundary representation with edges as boundary elements.

- Vector-Boundary Representation (VBR)

With the vector-boundary representation arbitrary polyehedra can be represented. Complex geometries are described with the hierarchy "geometry → polygons→ edges → points". Fig. 2.3 shows a geological fault represented in the vector-boundary representation.

---

1. For the geoscientific reader we emphasize that the following representations of a geological fault are abstracting the geometries as they occur in nature. In so far they are "unrealistic" according to the "real" geometries that are the consequence of geological processes.
2. The second case will be treated in chapter 2.2.3.

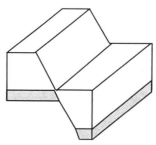

Fig. 2.3 Geological fault (push down of the right block along a sloped plain) represented in the vector-boundary
representation[1]

Advantages of the vector-boundary representations are:

- geometric and topological information can be clearly separated. Thus, for transformations of the coordinates only the point coordinates have to be updated. The transformations let the topology of the objects be unchanged. Furhermore, the topological
  consistency can be easily checked,

- the description of the geometry borders (boundary) is explicitly given.

Disadvantages of the vector-boundary representation are:

- topological relationships and information about solids are not existing,

- set operations like intersection tests are costly to compute, i.e. complex algorithms are
  necessary,

- geometries with curved surfaces must be approximated.

The application of the vector-boundary representation is limited. We can use it for surface
representations in a 3D-GIS, in connection with the function-boundary representation,
which will be introduced below. It can be seen as a direct extension of the 2D-vector representation.

2.2.2 Tesselating Representations

The tesselating representations divide the 3D space into a set of adjacent, i.e. not overlapping, primitive geometries. These primitives can be of different types.

---

1. The grey underlayed layer shall emphasize the geological process which is behind the picture. The below
border is to be seen only as a border for the representation.

- Cell-Decomposition

With the (irregular) cell-decomposition complex objects can be composed of simple building blocks ("cells") like cubes, cuboids, cylinders or tetrahedra according to the "principle of a toolbox". The composing then corresponds to a union that does not allow intersections. Different shapes and sizes of cells are admitted. Thus not all of the cells must be identical. Fig. 2.4 shows an example of a cell-decomposition of two cell types for the modeling of a geological fault. An important application of the cell-decomposition is the finite element method (FEM).

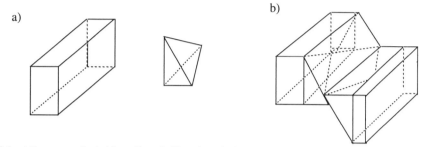

Fig. 2.4   a) Two types of primitive cells: cuboids and tetrahedra
            b) Example of a cell-decomposition of a geological fault with cuboids and tetrahedra

Advantages of the cell-decomposition are:

   - Accurate approximations for arbitrary solids are provided,
   - free-form areas can be sufficiently represented.

Disadvantages of the cell-decompositions are:

   - The complexity of geometric algorithms is depending on the primitive cells,
   - there is no information about the topology of the primitive cells.

- Spatial Occupancy Enumeration

Spatial occupancy enumeration (see fig. 2.5) is a special case of the cell-decomposition. The geometry is decomposed into identical cells (cubes) that are organized in a fixed, regular grid in space. Space is unambigiously defined, i.e. it cannot be occupied by two or more geometries. The cells are adjacent and connected, i.e. they are not overlapping. Each cell is adressable by an unambigious code. The reduction of the spatial occupancy enumeration to 2D space is equivalent to the raster representation.

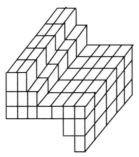

Fig. 2.5 A geological fault represented with the spatial occupancy enumeration

Advantages of the spatial occupancy enumerations:

- cells are spatially indexed, i.e. they are adressable, which allows an efficient search of the geometries,
- the determination of intersection, union and difference of two objects is simple,
- topological properties like neighbourhood, adjacency etc. are implicitly given and thus they can be easily computed.

Disadvantages of the spatial occupancy enumeration are:

- For many geometries the representation is only an inaccurate approximation, as the solution of the enumeration cannot be infinitely by definition,
- there are no partially occupied cells, i.e. unfortunately "empty space" is managed,
- the geometries are transformation- and rotation-invariant, i.e. object transformations, such as translation and rotation are very costly,
- the storage place to be needed for a good approximation is high.

One of the well known applications of the spatial occupancy enumeration is the computer-axial-tomography (CAT) in medicine. An example are *octrees* (SAMET 1990a, 1990b), a hierarchical variant of the spatial occupancy enumeration with the goal of saving storage space. Hereby the geometry is binary divided with the divide-and-conquer method. An octand can be "filled", "partially filled" or "empty", depending on how far the geometry intersects the octand (see fig. 2.6). A partially filled octand is recursively divided in sub-octands. The subdivision ends in its last level, if all octands are either empty or filled, or if the a priori defined depth of the octree is achieved. The subdivisions may be represented in a tree.

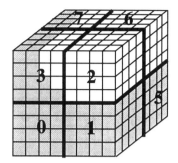

Fig. 2.6 Octree enumeration with the example of a geological fault.

Today octrees are a wide-spread spatial representation in the geosciences. They are developed from the voxel representation, which is corresponding to the raster representation in 2D space. The wide spreading of raster data and their straight forward extension in the third dimension (voxel) seems to make the octree to an interesting alternative for a 3D-GIS. However, depending on the space of the geometric data and their location, the octree may be very insufficient because of its large need for storage place. This can be easily seen with the example of large areas in 3D space. Tests like those of GUNTERMANN (1994) confirm that efficient intersection operations can be easily realized on the octree, but for the computation of surfaces a high uncertainty must be accepted.

2.2.3 Analytical Representations

We also give the known analytical representations in 3D space for completeness. By an analytical representation we understand a representation with which free-form areas (e.g. spline- or Bezier-areas) can be represented by functions and parameters. We first list the second case of the boundary representation mentioned in chapter 2.2.1.

- Function-Boundary Representation (FBR)

We speak of the *function-boundary representation (FBR),* if the bounding elements of the objects are analytical functions or areas defined by interpolation- or approximation-methods. Fig. 2.7 shows a geological fault represented in the function-boundary representation.

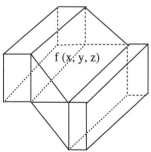

Fig. 2.7 Geological fault given in the function-boundary representation

The function-boundary representation, however, has the big advantage that geometric properties are difficult to check and often demand costly algorithms.

- Sweep Representation

The geometries of the objects are represented by rotation around an axis *(rotation sweep)*, by translation of a line *(translation sweep)* or by a combination of sweeps *(general sweep)*. Fig. 2.8 schematically shows an example for the geometry of a geological fault represented by a translation sweep. The parts *A, B* and *C* were generated by a translation sweep along the y-axis of the thick framed geometries.

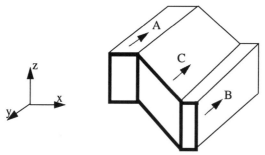

Fig. 2.8 Geological fault represented by a translation sweep

The sweep representations are particularly suited for fixed free-form areas, as they appear in the production technology, for instance. Only the general sweep, however, has a sufficient domain for the description of arbitrary geometries. Furthermore, the sweep representation is no closed system, i.e. the union, for example, of two geometries represented with a translation sweep need not necessarily be again a sweep representable geometry. Unfortunately, the computation of geometric operations is also very costly. These disadvantages seem to be a too big limitation for the application in a 3D-GIS.

- Primitive instancing (parameter representation)

By the primitive instance, the geometries of the objects are described by a fixed number of *parameters*. An *instance* of a primitive is defined by a set of numerical values. Each value is defined in the mathematical equation that describes a solid. Primitive instancing is the best suited representation for the composing of regular geometries.

With the parameter representation we can easily model a priori fixed families of geometries. Each element of the family is characterized with a fixed number of parameters. Fig. 2.9 shows a geological fault represented in the parameter representation.

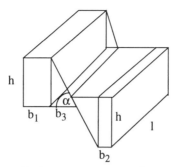

Fig. 2.9 Geological fault represented in the parameter representation

Primitive instancing is particularly suited for the construction of variants in CAD applications, as only the parameters must be changed for the generation of new variants. However, fixed variants play only a minor role for GIS. A disadvantage is the limited domain. More complex geometries cannot easily be hierarchically composed with primitives like cubes etc. In contrast to that complex geometries are modeled with loose parameters that are composed by a set of parameters. Altogether the parameter representation is little "tailor-made" for the geometric modeling within a 3D-GIS.

2.2.4 Hybrid Representations

- Constructive Solid Geometry (CSG)

In the so called construction solid geometry representation the geometries of the objects are represented by the (recursive) binary *composition of primitive solids* (e.g. cubes, cones, cylinders, torus etc.). Only the set operations (union, intersection, difference) and transformations (translation, rotation and scaling) are allowed, i.e. the geometry of an object can be defined as a set theoretical combination of standard primitives. The construction solid geometry can be defined recursively by a grammar. Every object is representable as a binary tree, where the leaves correspond to the primitive solids and the nodes to the operations, i.e. the set operations and the transformations. The construction tree (see fig. 2.10) is the graphical representation of a boolean expression over primitives. As the boolean operations are not commutative in the general case, the edges of the tree have to be sorted. The primitives themselves can be represented in an arbitrary representation like primitive instancing. In the POLY-system (MEIER and LOACKER 1987), for example, a combined representation with constructive solid geometry and the boundary representation is used. The same is true for the implementation of the geometric data in AIM-P described in (DYBALLA et al. 1991). In this system the geometries are only accessable by CSG operations, whereas internally in the database they are managed by a boundary representation. Thus the modeling with constructive solid geometry is situated a level "higher" than the other representations. We can also consider the cell decomposition and the spatial occupancy enumeration as special cases of the constructive solid geometry representation, if the only allowed operator is the composing of geometries. Then the interiors of the geometries must not intersect.

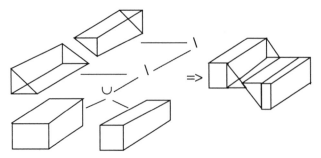

Fig. 2.10 Geological fault represented by constructive solid geometry

An advantage of the constructive solid geometry representation is that the generation of the geometries takes place interactively by a simple modeling language and a graphical construction tree. A disadvantage is that the representation by a construction tree is not unambigious, i.e. two different representations may describe the same geometry. Unfortunately, free-form areas are difficult to model by construction trees. Furthermore, even small changes of the primitives usually demand totally new and complex algorithms. Because of the awkward representation of free-form areas the CSG representation is not suited for a 3D-GIS. Besides the introduced spatial representations experimental, combinations of different representations may be used. The PM octree[1] (SAMET 1990a), for example, is a combination of the octree with the boundary representation. In the PM octree the octree is recursively subdivided into nodes, until it ends in one of five possible types of leaf nodes. The leaf nodes may be filled or empty (like in the "usual" octree) or they may consist of three additional types for partially filled nodes, which are called point nodes, edge nodes and area nodes, respectively. Point nodes contain points with their connected areas. Edge nodes contain (a part of) an edge with its connected areas. Finally, area nodes contain the clipped part of an area, which is inside the corresponding octant of the octree.

2.3 Comparison of the Representations

We will now try to compare the different 3D-representations to discuss their suitability for a 3D-GIS. For the comparison the following criteria are used[2]:

1) Domain
   Which objects can be represented? Are there any limitations?

2) Validity
   a) geometric validity

   Are all of the objects valid objects, i.e. are they finitely representable and do they fulfil the a priori defined geometric and topological constraints of the respective representation (e.g. for the boundary representation: "Areas and edges may only intersect in common edges or points")?

---

1. PM stands for **Polygonal Map**
2. The first four criteria (1, 2a, 3 and 4) were already mentioned by MÄNTYLA (1988).

b) "technical" validity

Independent of the geometric validity, for the user the so called "technical" validity has to be considered. For example objects represented in the sweep representation (especially rotation sweep) may lead to technically wrong representations[1]. The geologist, for instance, has his checking criteria in mind that tell him what is "correct" and consistent, respectively. He orientates himself by means of geological processes or hypothesis.

3) Non-ambiguity and unambiguity

Is there exactly one object for each representation and exactly one representation for each object?

4) Closed operations

Do the operations on the represented objects again lead to objects of the same representation?

5) Efficiency of the geometric algorithms

Canthe runtime of geometric algorithms be well estimated?

6) Accuracy

How accurate is the representation of the objects in comparison with the objects of the "real world"?

7) Need for storage

How big is the need for storage for the objects and what does it depend from?

Table 2.2 shows a comparison of the different representations according to the seven criteria.

---

1. In delimination from the terms *accuracy* and *precision* we note that, for example, a voxel representation may be correct, but inaccurate. In contrast to that a rotation sweep can be accurate, but technically wrong.

Representation → _____ Criterion ↓	Edge model	Sweep-Representation	Parameter Representation	CSG	Boundary Repr. (VBR/FBR)	Spatial occupancy enumeration	Cell decomposition
Domain	none ~1	none ~2	none ~3	dep. ~4	only ~5	all ~6	all ~6
Validity - geometric - technical	no no	yes no	yes yes	yes yes	yes yes	yes yes	yes yes
Non-ambiguity and unambiguity	no no	yes no	yes yes	yes no	yes no	yes yes	yes no
Closed operations	no	yes	yes	yes	yes	yes	yes
Efficiency	+	-	-	-	+/-	+	+
Accuracy	-	++	++	++	-/++	-	+
Need for storage	low	very low	very low	very low	high/ low	very high	very high
Suitability for 3D-GIS	- -	-	-	-	(+)	(+)	(+)

[1] areas and volumes; [2] irregular objects; [3] free-form areas, [4] dependent on primitives; [5] (sur-)faces, no volumes; [6] approximated objects.

Table 2.2 Comparison of 3D-representations

Obviously the edge model is ruled out for the application in a 3D-GIS, as we can generate representations that are no valid objects and because several objects may exist to a single re-presentation. Furthermore, any area- and volume information is missing. For the sweep-, the parameter- and the CSG-representation the domain is too limited. Furthermore, a sweep re-presentation may lead to technically wrong geometries especially in the rotation sweep. The CSG-representation is at best suited for the visualization of simple 3D-geometries because of its limited number of primitives. Geometric CSG operations like volume- and surface-computations can only be executed by means of a conversion, for instance into the spatial occupancy enumeration. However, this leads either to inaccurate results or to innacceptable runtimes (compare e.g. IFFLAND 1994). As the volume information, which is urgently needed in a 3D-GIS, is missing to the two boundary representations VBR and FBR, they are only well suited for the representation of (sur-)faces. Most of the introduced representations are not unambigious. However, this is no exclusion criterion. During the interactive modeling of geologically defined geometries (see chapter 7), for example, different versions of the same geometry that are generated at different times, ar even desired.

We can subdivide the representations into such representations that model accurate geometries and in such that provide a more or less good approximation. Comparing the spatial occupancy enumeration and the cell decomposition, the spatial occupancy enumeration significantly provides the worse approximation. The domain of the spatial occupancy enumeration is dependent on the size and the shape of single cells in space. For the octree representation, which is a special case of the spatial occupancy enumeration, we emphasize the efficient realization of geometric operations. Examples are the intersection of objects or the spatial search (BAK and MILL 1989; GUNTERMANN 1994). In the main this is also true for the cell decomposition. Thus the boundary representation, the spatial occupancy enumeration and the cell decomposition seem to be the most suited of the examined representations for the modeling and management of geo-objects in a 3D-GIS. Likewise, from a geoscientific point of view, these three representations map geometries of the "real world" into the spatial representation in such a way that they correspond closely to their shape in nature.

2.4 Conversions

The current way for the integration of spatial representations in GISs is the conversion from the vector into the raster representation and vice versa. In the following we will discuss the problems occuring during the conversion.

<u>Conversion between 2D-representations</u>:

Converting the vector into raster data is usually called a "rastering" of the vector representation. For the way back we say that the raster representation is "vectorized". The conversion of vectors into raster cells is easily enabled by putting a raster over the vectors and by marking those raster cells that are filled by a vector. The conversion of classified raster, however, into a set of homogeneous polygons is many times more costly (see ABEL and WILSON 1990; ZHOU and GARNER 1991). Depending on the method used and the shape of the geometries, serious approximation errors can arise. Thus considerable information losses can occur during the conversion of data between raster and vector representations and vice versa.

Theorem 2.1: Let *ov: O → V and or: O → R* be mappings[1] between the object space and the vector and raster space, respectively. Let *vr : V → R and rv : R → V* be mappings for the conversion between vector and raster geometries. Then:

(1) An o ∈ O may exist with rv (or (o)) ≠ ov (o) ∧
(2) an o ∈ O may exist with vr (ov (o)) ≠ o r (o).

Proof:

Part (1):   Assumption: ∀ o ∈ O: rv (or (o)) = ov (o)

Let o_{11}, o_{12} be objects with a line geometry, so that the geometry of o_{11} is equal to the main diagonal of an arbitrary raster cell R, running from the left bottom to the right top corner. Let the geometry of o_{12} be the other main diagonal, running from the left top to the right bottom corner (see fig. 2.11).

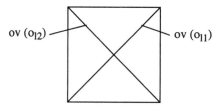

Fig. 2.11 Example of two line objects with the same raster representation

⇒ (rv (or (o_{11})) = ov (o_{11})) ∧ (rv (or (o_{12})) = ov (o_{12}))
⇒ (rv (R) = ov (o_{11})) ∧ (rv (R) = ov (o_{12}))
⇒ ov (o_{11}) = ov (o_{12})

Part (2):   Assumption: ∀ o ∈ O: vr (ov (o)) = or (o)

Let o ∈ O be an object with a circle geometry that covers two raster cells, let the second, however, be covered only little;
⇒ an ov (o) may exist so that ov (o) covers only a single raster cell (because of an approximation error);
⇒ vr (ov (o) covers a single cell and or (o) covers two cells;
⇒ vr (ov (o)) ≠ or (o).

The information loss from part (1) is drawn striped in fig. 2.12. If a raster is converted into vectors, generally the result is unequal to the vector representation being directly mapped from the object space. This is because the raster has a limited solution. The reversed mapping, i.e. the rastering of vectors, however, is equal to the raster representation that is directly mapped from the object space. Usually in this case the information loss is smaller as it only consists of the difference between the geometry of the object in the "real world" and its ap-

---

1. Following MEIER (1986), let *O* be the "object space", i.e. the set of geometric objects. Let *V* and *R* be examples for the "representation space", which contains the set of syntactically and semantically correct representations. *ov* and *or* are mapping the object space *O* into the vector space *V* and the raster space *R*, respectively.

proximation in the vector representation. In the example of fig. 2.1, the conversion of the geometry of Berlin, the information loss from the vector to the raster representation can be seen very clearly, as the size of raster cells are sketched exaggerated, i.e. very inaccurately.

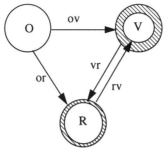

Fig. 2.12 Information losses for the conversion between raster and vector representations

The conversion problems also lead to conceptual difficulties. Vector based systems (GISs) expect a high accuracy of the geometry values. However, this accuracy cannot be provided by raster systems. Raster systems provide a vast amount of data that mostly cannot be managed in vector-based systems. Extremely bothering are also effects that arise during the back transformation. Because of the inaccuracy that arises after the mapping from real numbers (vectors) upon the raster, the two following mappings

(3) $rr := rv(vr(v))$ and $vv := vr(rv(r))$, $\forall\ r \in R,\ v \in V$

do not lead to the same result, in the general case. The "partwise" rotation into the raster representation that usually does not lead to the accurate starting situation after a rotation of 360 degrees, leads to a comparable unpleasant result. Let α and β be divisors of 360, then an $r \in R$ may exist such that:

(4) $rotation_{360}(r) \neq \alpha * rotation_\beta(r)$.

Conversion between 3D-representations

The quality of the conversion between different 2D-representations is of great importance for their use in a 3D-GIS. As we will see, however, the problems for the conversion between 3D-representations are even graver. A comparable effortless conversion is that from the boundary representation to the linear octree coding (TAMMINEN and SAMET 1984; TAMMINEN et al. 1984). An overview of methods for the conversion between 3D-representations is given by (REQUICHA and VOELCKER 1983). Table 2.3 gives an actualized version with completions by (WEICK 1988) and (MÄNTYLA 1988).

to → ―――――― from ↓	Sweep Repr.	Cell Decomposition	CSG	FBR	VBR	Spatial Occupancy Enum.
Sweep Repr.		K	K	K	K	K
Cell Decompos.	U		K (U)	K	K	K
CSG	U	K		K	K	K
FBR	U	K	U		K	K
VBR	U	K	U	U		K
Spatial Occupancy Enumeration	U	U	U	U	K	

K: Known from literature; U: Unknown

Table 2.3 Conversions between spatial representations in 3D space

We respectively direct our attention to table 2.3 column by column, i.e. to the conversion of the representation listed in the left column to those listed in the head line of the table. Unknown or at best known for limited application fields are the conversions to the sweep representation. Cell decompositions can be seen as a limited version of CSG, but algorithms for the generation of general CSG from cell decompositions are not known to the author. From the vector-boundary representation (VBR) and from the spatial occupancy enumeration to the function-boundary representation (FBR) no conversion algorithms are known to us.

Because of the conversion problems, the sweep representation, CSG and FBR are disqualified for a unified representation in a 3D-GIS. Thus from our previous "candidates" only the cell decomposition, VBR and the spatial occupancy enumeration are left over. In the VBR, however, the volume information is missing. The cell decomposition provides no internal topology and the spatial occupancy enumeration has the disadvantage of an inaccurate approximation.

A large problem for the conversion of spatial representations is the *consistency* of different representations. Hitherto no hybrid tools are known that can update different representations of the same object, as for example a CSG and a boundary representation, simultaneously. Thus the consistency of the different representations is not ensured.

ERRATA

Lecture Notes in Earth Sciences, Vol. 61
M. Breunig, Integration of Spatial Information
for Geo-Information Systems
(ISBN 3-540-60856-7)

Unfortunately in print pages 33 and 41 have been falsely printed so that the text that should be on page 41 is on page 33 and vice versa.
We apologize for this error.

Chapter 4

A Model for the Integration of Spatial Information

At the beginning of this chapter a classification of spatial operations[1] is introduced. It will be the basis of a "three-level notion of space" for the integration of spatial information. Following the relational algebra, so called "building blocks" of spatial queries for GISs are introduced, in which spatial relationships and operators can be embedded. As the basis of our model we introduce the *extended complex (e-complex)*. It provides the explicit representation and management of the geometry and topology of geo-objects. Particularly for the representation of three-dimensional surfaces and volumes, we transfer the e-complex into a convex e-complex *(ce-complex)*. Thus the space between the e-complex and the convex hull is filled with "virtual tetrahedra". As we know from computational geometry, geometric algorithms on convex objects are less complex. They additionally gain in efficiency by the explicit use of the topology of the ce-complex. We give a short analysis for the transformation of other spatial representations into the e-complex. The so called "ECOM-algebra" is characterized by topological, metrical and direction relationships and operators on e-complexes. Furthermore, special topological relationships are introduced that are defined on the internal topology of the e-complexes. Finally the extension to set-valued operations allows the treatment of non-connecting resulting objects.

4.1 Classification of Spatial Operations

We first subdivide the spatial operations into *spatial relationships* and *spatial operators,* depending on their result, i.e., the spatial operations provide a result of type boolean or of the type "geo-object". EGENHOFER (1989, 1989a, 1989b, 1991; KAINZ 1991) and other authors already proposed a convenient classification of binary spatial relationships into topological (sometimes also called set relationships or boolean relationships), metrical and order or direction relationships. For the last group of relationships the order of the operands is decisive[2], i.e. $A \ OrderRel \ B \Leftrightarrow B \ OrderRel^{Inv} \ A$. We extend the validity of this classification for general, i.e. non-binary spatial relationships and for spatial operators. Table 4.1 shows a classification of spatial operations and gives the most important examples, without a claim on completeness.

---

1. As we will see in the following, spatial operations include spatial relationships and spatial operators.
2. Let $OrderRel^{Inv}$ be the inverse relationship of OrderRel. *RightOf,* for instance, is the inverse relationship of *LeftOf.*

3.1.2 Estimations

We just name shortly another kind of approximation, namely the estimation of the geometry by spline- and Bezier-areas. In a similiar way as in the bounding-box world, we abstract from the real coordinates, i.e. from the geometry. Its place is taken from a non-trivial parametrized description of the approximation, which is computed by a set of given points.

Spatial relationships on spline- and Bezier-areas

Fortunately spatial relationships between geometries being approximated with polynoms can be computed by the solution of equation systems. However, in the general case, very costly computations are necessary.

3.2 The Point-Set Approach

In the so called point-set approach the geometries of the geo-objects are defined by their infinite point sets. By that an abstraction from the concrete geometric realization can be managed without accepting the disadvantages of an approximation.

Spatial relationships on point sets

Topological relationships like equality, inclusion and intersection can be defined on point sets by the usual set operators. Following GÜTING (1988, 1988a) we give some examples:

geom (a)	=	*geom (b)*	:= *points (a)*	=	*points (b)*	
geom (a)	≠	*geom (b)*	:= *points (a)*	≠	*points (b)*	
geom (a)	*inside*	*geom (b)*	:= *points (a)*	⊆	*points (b)*	
geom (a)	*outside*	*geom (b)*	:= *points (a)*	∩	*points (b)*	= ∅
geom (a)	*intersects*	*geom (b)*	:= *points (a)*	∩	*points (b)*	≠ ∅
geom (a)	*neighbour*	*geom (b)*	:= *points (a)*	∩	*points (b)*	≠ ∅

Metrical relationships between point sets (distances) can be executed on labeled points, as for instance the centre of gravity. Thus the usual distance measures like the Euclidean distance can be used.

A lack of the point set approach is that the intersection and the neighbourhood relationship base on the same definition (EGENHOFER 1989). That is why the point-set approach was extended by the differentiation of boundary and interior (PULLAR 1988). Thus an unambigious definition of the neighbourhood relationship is possible. A disadvantage is that the point-set approach favours the raster representation, because every geometry can be represented as a set of pixels. As the point-set approach presupposes an infinite geometry, it cannot be directly implemented in the computer.

3.3 Topological Abstractions

With the help of "topological abstractions" which we introduce in the following, an important class of spatial relationships on such objects can be specified that are similiar to the objects of the "real world". The topological relationships come next to the human understanding of the position of objects in space, which we apparantly assess it daily with the greatest of ease. The description takes place on a high abstraction level without consideration of concrete coordinates.

3.3.1 The "Blocks World"

In the "blocks world", originally derived from the childrens world, geometries of objects can be simply defined by "blocks" that are lying on a table.

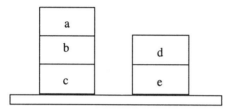

Fig. 3.1 Geometric objects in the blocks world

In fig. 3.1 the set of geometric objects is {a, b, c, d, e}. Between the objects the following spatial relationships can be defined.

Spatial relationships in the blocks world

An example of a topological relationship in the blocks world is the relationship "upon" that is true between two blocks, iff a first block is directly lying upon a second block, i.e. if both blocks are neighbouring. In our example the relationship *"upon"* is true for the following set of tuples *{(a, b), (b, c), (d, e)}*. Correspondingly a relationship *"above"* may be introduced that is true between two blocks, iff the first is (somewhere) above the second.

An advantage of the blocks world is that simple and general theorems can be formulated for the spatial relationships between objects, for example in the predicate calculus. We give the following two examples:

1) If in the blocks world a first block is directly lying *upon* a second block, then the first block also lies *above* the second block:

$\forall x \, \forall y \; (upon \, (x, y) \; \Rightarrow \; above \, (x, y))$.

2) The relationship *above* is transitive. If a first block lies above a second one, and the second lies above a third one, then the first also lies above the third:

$\forall x \, \forall y \; (above \, (x, y) \wedge above \, (y, z) \; \Rightarrow \; above \, (x, z))$

The same is true for the upon-relationship. Thus "object x lies above object z" can be defined in the following way:

above (x, z) ⇔ *upon (x, z)*
$$\vee\ (\exists\ y_1, ..., y_n : upon\ (y_n, z) \wedge upon\ (y_{n-1}, y_n) \wedge ... \wedge upon\ (x, y_1)).$$

The disadvantage of the blocks world is the small set of spatial relationships which can be defined. *Metrical relationships* and *direction relationships* between arbitrary placed objects[1] in space are not defined in the blocks world.

3.3.2 Cell Complexes

PIGOT (1992) introduced a topological model that is based on so called *manifolds, cells* and *cell-complexes*. In this model, depending on the dimension of space, the following cells are defined:

0-cell ≡ point,
1-cell ≡ edge,
2-cell ≡ circle,
3-cell ≡ sphere,
4-cell ≡ sphere at a given time.

A k-dimensional manifold is a *topological space*[2] *X,* in which every point $x \in X$ has a neighbourhood to a k-dimensional environment. A *k-cell* is a bordered subset of a k-manifold. *A k-cell complex* is the union of all k-dimensional and lower dimensional cells.

Spatial relationships on cell-complexes

The most important operators on cell-complexes are the *boundary-* and the *coboundary-operator*. The boundary-operator provides the set of the (k-1)-cells for a k-cell, i.e. the cells of dimension (k-1) the k-cell consists of. The coboundary-operator provides the corresponding (k+1)-cells for a k-cell.

Fig. 3.2 shows an example of a 2-cell for both operators. The result is drawn black, respectively. In example a) the result is the boundary which borders the area of the circle *C* and in b) the result is the sphere that borders the area of the circle *C.*.

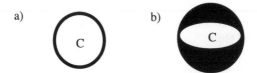

Fig. 3.2 a) Boundary of a 2-cell C
        b) Coboundary of a 2-cell C

---

1. Examples for direction relationships are *westOf/eastOf* etc. We will introduce the different types
   of spatial relationships in chapter 4.1.
2. The topological space will be formally introduced in chapter 4.3.

In the model of Pigot traversion operators (so called local operators, i.e. defined on a single cell complex), identification operators (so called global operators, i.e. defined between different cell-complexes) and construction operators (operators of a lower level) are defined on cell complexes.

Let us now have a closer look on a special case of cell-complexes, the so called *simplicial complexes,* which were especially examined by EGENHOFER (1989, 1989a, 1989b et al. 1989).

3.3.3 Simplicial Complexes

Simplices (MOISE 1977; EGENHOFER 1989b) can be topologically classified towards their dimension in the following way:

0 - simplex ≡ node,
1 - simplex ≡ edge,
2 - simplex ≡ triangle,
3 - simplex ≡ tetrahedron.

The following definition is taken from (DTV 1991).

Definition 3.1: Let $P_0, .., P_s$ be independent points from $\Re_n$ ($s \le n$), then $\sigma^s := \{P \mid \lambda_0 P_0 + ... + \lambda_s P_s \wedge \lambda_i \in R_0^+ \wedge \lambda_0 + ... + \lambda_s = 1\}$ is called *s-dimensional geometric simplex.*

It is interesting that simplices again are composed of simplices of a lower dimension. In analogy to the term "faces of a triangle" we speak of face simplices of a simplex (DTV 1991), and faces of a simplex, for shortness.

In the n-dimensional space a n-simplex has $\binom{n+1}{d+1}$ faces of the dimension d. Thus in the 3-dimensional space a simplex has $\binom{3+1}{0+1}$, i.e. 4 nodes, $\binom{3+1}{1+1}$, i.e. 6 edges, $\binom{3+1}{2+1}$, i.e. 4 triangles and $\binom{3+1}{3+1}$, i.e. 1 tetrahedron as faces (see also fig. 3.3). Thus a 4-dimensional simplex has 5 nodes, 10 edges, 10 triangles and 5 tetrahedra as faces.

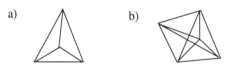

a)   b)

Fig. 3.3 a) 3-dim. Simplex b) 4-dim. Simplex

Definition 3.2: A *simplicial complex* is a finite set of simplices with the following properties:

(1) The intersection of two respective simplices is either empty or a face of both simplices,

(2) with every simplex each of its faces is also defined.

Fig. 3.4 shows an example of a simplicial complex.

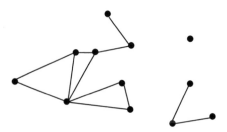

Fig. 3.4 Example of a simplicial complex

A simplicial 3-complex is a set of connected 3-simplices. Each 2-simplex is bordered by at most two 3-simplices and every 1-simplex is surrounded by at most two 2-simplices. Every 1-simplex connects two 0-simplices and the boundary of a simplicial 3-complex is a connected sequence of 2-simplices.

The dimension d of a simplicial complex C is defined as the maximal dimension of the simplices in C.

Theorem 3.1: Let C be a simplicial complex of dimension d $(d > 0)$, then the boundary of C, i.e. @C is a simplicial complex of dimension $(d-1)$.

<u>Proof</u>:
<u>Assumption</u>:   1) Let the boundary of C not be a simplicial complex.
                    $\Rightarrow \exists$ at least one face of C that is no simplex.
                    $\Rightarrow C$ is not a simplicial complex (contradiction).

      2)   Let the boundary of C be a simplicial complex that is not of the dimension $(d-1)$.
      2.1) Let the dimension of the boundary be smaller than $(d-1)$.
           $\Rightarrow$ for $d = 1$ is true: The dimension of the boundary is smaller than 0 (contradiction).
      2.2) Let the dimension of the boundary be greater than $(d-1)$.
           2.2.1) Let the dimension of the boundary be greater than d.
                $\Rightarrow \exists$ simplices of C with a dimension that is greater than d (contradiction).
           2.2.2) Let the dimension of the boundary be exactly d $(d > 1)$.
                $\Rightarrow$ In d-dimensional space is true: $\exists$ points P_i on the boundary of C for whose environments is true: $\forall P_{ui} \in$ environment of P_i: $P_{ui} \in$ K.
                $\Rightarrow P_i$ is not a boundary point (contradiction).$\blacklozenge$

An important advantage of the simplices is their simple structure. In every dimension they consist of the simplest geometry, respectively.

Spatial relationships with simplices

EGENHOFER (1989, 1989a, 1989b) proposes to use simplices for the definition of binary topological relationships. Topological relationships are based upon the concept of topological space and the terms *"interior", "hull"* and *"boundary"* of open sets.

The intersection of the interior with the boundary of an open set is empty and the union of the interior and the boundary is the hull of the set.

The approach is based upon the intersection of the boundary and the interior of two topologies. It is differentiated between empty and not empty results of intersection. Elementary operations on simplices are the boundary operation and the complementary interior-operator.

Whereas the *boundary-operator* provides a set of (k-1)-simplices for a k-simplex, the *coboundary-operator* has the neighbouring (k+1)-simplices of the k-simplex as a result. Fig. 3.5 shows the boundary- and the coboundary-operator for a 2-simplex. The result is drawn with a fat black line, respectively.

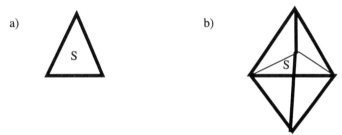

Fig. 3.5 a) Boundary of a 2-simplex S; b) Coboundary of a 2-simplex S

In fig. 3.5 a) the result consists of three 1-simplices, which build a triangle. In fig. 3.5b) the both tetrahedra (3-simplices) which border to *S* are the result of the coboundary-operator.

EGENHOFER (1989a, 1989b) showed that the following *minimal set of topological relationships* between n-simplices (n > 0) can be determined by the four possible intersections (in 2-dimensional space) of the boundaries and the interiors of two spatial objects *A* and *B*:

$$@S_1 \cap @S_2, \quad {}^{o}S_1 \cap {}^{o}S_2, \quad {}^{o}S_1 \cap @S_2, \quad @S_1 \cap {}^{o}S_2 \quad \text{(with } @ \equiv \text{Rand, } {}^{o} \equiv \text{Interior)}$$

The criterion for the differentiation are empty ($\varnothing$) and non-empty ($\neg\varnothing$) intersections.

S_1 S_2	$@ \cap @$	$° \cap °$	$@ \cap °$	$° \cap @$
disjoint	$\varnothing$	$\varnothing$	$\varnothing$	$\varnothing$
meet	$\neg\varnothing$	$\varnothing$	$\varnothing$	$\varnothing$
overlap	$\neg\varnothing$	$\neg\varnothing$	$\neg\varnothing$	$\neg\varnothing$
covers	$\neg\varnothing$	$\neg\varnothing$	$\varnothing$	$\neg\varnothing$
coveredBy	$\neg\varnothing$	$\neg\varnothing$	$\neg\varnothing$	$\varnothing$
inside	$\varnothing$	$\neg\varnothing$	$\neg\varnothing$	$\varnothing$
contains	$\varnothing$	$\neg\varnothing$	$\varnothing$	$\neg\varnothing$
equal	$\neg\varnothing$	$\neg\varnothing$	$\varnothing$	$\varnothing$

Table 3.1 The specification of a minimal set of binary topological relationships on n-simplices ($n > 0$, compare EGENHOFER (1989b), based on the criterion of empty and non-empty intersections of boundaries and interiors.

The *overlap*-relationship, for instance, provides the value "true" as result, if the boundaries of both complexes intersect themselves ($@ \cap @ \neg \varnothing$) or if boundaries and interiors intersect themselves (($@ \cap ° \neg \varnothing$) and ($° \cap @ \neg \varnothing$)), respectively or if the interiors of the complexes intersect themselves ($° \cap ° \neg \varnothing$). *covers/coveredBy* and *inside/contains* are redundant relationships, respectively. That is why *coveredBy* and *contains* may be taken out of table 3.1. Thus six topological relationships are left over.

A disadvantage of the simplicial complexes is that *direction relationships* like *above, below, left, right* etc. cannot be expressed. All these relationships are categorized as "outside". Thus they cannot be considered separately. Furthermore, *metrical relationships* are not defined, as the reasoning is explicitly based on topology.

3.4 Valuation

Approximations of geometries have the advantage that one can define all the spatial relationships needed for a 3D-GIS. However, for the determination of spatial relationships, we must accept a potential error which is dependent on the quality of the approximation. In the introduced approximation by cuboids this error is extremely large.

The *point-set approach* is interesting for theoretical considerations, but it cannot directly be transformed to an implementation for a spatial data model of a GIS.

Topological abstractions are an elegant possibilty to specify an important class of spatial relationships. Non topological relationships, however, cannot be defined. In the following it will be our goal to use the advantages of topological abstractions and to extend them with further concepts.

Chapter 3

Abstraction from the Spatial Representation

The problems of the conversion between different spatial representations are our motivation to examine another way for the integration of spatial information. In this chapter, approaches are introduced that can be used to specify spatial operations on geo-objects independently of their spatial representation. In this context we speak of an *"abstraction"* from the spatial representation. Thus a GIS-user has a transparent view upon the spatial representation of the geo-objects and the specification is excused from unnecessary details: Besides approximations, the "point-set approach" and so called topological abstractions are introduced that have the property to abstract from the concrete geometry in an elegant way, i.e. they abstract from the single coordinates of the geo-objects.

3.1 Approximations

3.1.1 The "Bounding-Box World"

As "bounding-box world" we designate a model that describes the geometry of geo-objects with their minimal circumscribing cuboids, the so called bounding boxes. In the simplest case the bounding boxes are oreientated towards the x-, y- and z-axes.

Spatial relationships in the bounding-box world

Spatial relationships are exclusively defined on the bounding boxes of the geometries. Thus arbitrary spatial relationships can be defined. A disadvantage is the usually bad approximation of the geometry by bounding boxes and the resulting high number of "false drops", i.e. spatial predicates in which the bounding boxes do qualify themselves, but the geometries do not. The intersection of two bounding boxes, for instance, does not include by any means that the geometries lying inside the bounding boxes are also intersecting themselves.

In the bounding-box world spatial relationships can be defined by an extension of the interval logic (1D) into the second and the third dimension (MUKERJEE 1989). For instance the relationships "Object A is situated before/behind, leftOf, rightOf, above/below etc. object B" can be experessed by means of intervals along the x-, y- and z-axes.

| | Type of the operation | | |
	Topological	**Metrical**	**Direction**
Binary spatial relationships: geo × geo → bool **Spatial relationships:** {geo} × {geo} → bool	neighbour-hood, intersection, equivalence, inclusion, exclusion.	distance, angle.	northOf, westOf, southOf, eastOf; aboveOf, belowOf; leftOf, rightOf; before, behind.
Unary spatial operators: geo → real	-	length, surface, volume;	-
Binary spatial operators: geo × geo → real geo × geo → {geo*} **Spatial operators:** {geo}×{geo} → {geo*}	intersection, union, difference.	distance[a], angle[a].	-

a. Exclusively binary operator.

Table 4.1 Classification of spatial operations.
{geo*} stands for a set of new generated geo-objects[1]

Spatial relationships like the *exclusion* or *northOf* analyse a predicate, whereas *spatial operators* provide a set of real numbers or a set of new generated or computed geo-objects as their results. *Unary* and *binary spatial operators* are special cases, for which only one or two geo-objects are allowed as input. *Topological relationships* are invariant against geometric transformations. I.e. they stay unchanged, if the relevant objects are translated, rotated or scaled. They describe the position between geo-objects, like "is a point inside a polygon?" or "do the two polygons intersect?" A special case of the topological relationships *intersection* and *inclusion* is the "window-predicate": *window x geo -> bool*, that tests, if a set of geo-objects is inside or intersects a window, i.e. a rectangular section of a map. The classical *topological operators* are intersection, union and difference. They provide new generated geo-objects as a result. *Metrical relationships* analyse a "metrical predicate". We give an example: "Is the distance between Bonn and Düsseldorf larger than 50 km?" provides the value *true*

---

1. With the difference-operation new objects are not generated in any case.

as result. Examples for *direction relationships* are *northOf, westOf, southOf, eastOf, aboveOf* and *belowOf.*

One could imagine topological operators that, for instance, would provide all the neighbouring objects for a geo-object. This, however, is nothing but the sequential analysis of the corresponding binary topological relationship, with a following collecting of the qualifying objects. Thus no new objects would be computed. That is why we renounce on the explicit introduction of such operators. The same holds for direction operators.

Examples of metrical operators are the length of a line (unary operator) or the distance between two points (binary operator). Metrical operators provide a real number as result.

4.2 Building Blocks for Spatial Queries

We introduce the elementary differentiation between GIS-queries on a single geo-object *("local queries")* and GIS-queries on several geo-objects *("global queries")*. Examples of local queries are "what is the volume of a single geo-object?" or "what is the neighbourhood of triangles within a single triangle network?". Global queries are likely to give an answer to the position, the intersection or to the distance between two geo-objects. Obviously, these kinds of queries must also be considered for the choice of the spatial representation of the geo-objects (chapter 4.4) and for the conception of spatial access methods (chapter 5.3). It is interesting that there is an analogy with the object-oriented modeling, as the difference between methods of a single object class on the one side and relationships between objects of different object classes on the other side, demand different concepts[1]. The question arises, if a single geo-object eventually should be treated differently than a geo-object within a *set* of geo-objects. This implicates the question, if the differentiation between global and local queries should influence the spatial representation of the geo-objects. We will take up again this question in chapter 4.4. Let us first devote to the composition of building blocks for spatial queries. We distinguish between two kinds of building blocks, namely *basic building blocks* and *advanced building blocks*.

4.2.1 Basic Building Blocks

Following the terminology of the relational algebra, we introduce basic building blocks for spatial queries. The most important basic building blocks are spatial predicates and the already introduced *spatial operators*. A spatial relationship is a special case of a spatial predicate. Let us introduce the terms *spatial attribute* and *spatial predicate*.

A *spatial attribute* describes the topology or the geometry of a geo-object. Examples of spatial attributes are the topological description of a city with its districts or the point sequence of the geometry of a river.

---

1. We refer to "relationship objects", which are not yet realized in today's object-oriented programming languages, but which are internally realized as pointers within the objects.

A *spatial predicate* is characterized by the fact that its domain is the set of the domain of a spatial attribute.

With the basic building blocks, however, no complete spatial queries can be formulated. Thus we introduce so called *advanced building blocks.*

4.2.2 Advanced Building Blocks

The basic building blocks of spatial queries can be embedded into advanced building blocks. We distinguish between the *spatial selection,* the *spatial projection* and the *spatial join.* The building blocks can also be coupled with standard building blocks of the relational algebra without spatial reference like the selection or the projection (see example 3 below).

By a *spatial selection* we understand a selection of geo-objects from a geo-object set that is specified by one or several spatial predicates. A *spatial projection* is a projection of attributes of a geo-object with the condition that at least one spatial attribute is to be projected.

The association of geo-objects from two geo-object sets executed by the comparison of one of their spatial attributes, we call *spatial join.*

Example: In GISs so called *window queries* are often used. They are a special case of a spatial selection and select all geo-objects in 2D space that are inside or intersect a "window", i.e. a rectangle being directed to the x- and y-axes.

Let us give some examples of advanced building blocks in a graphical[1] and a SQL-like notaton:

1)

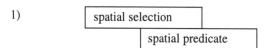

1a)   "Find all geo-objects, whose volume is larger than 10 m^3".

       SELECT *
          FROM geo-objects
             WHERE volume > 10 m^3

1b)   "Find all geo-objects, whose geometry, i.e. their coordinates, are inside the geo-
       metry specified by *geom*".

       SELECT *
          FROM geo-objects G
             WHERE (**inside** geom G.geometry)

___

1. Let the lower building block in the following figures be embedded in the upper(s), respectively. An arbitrary predicate, for instance, can be embedded in a spatial selection.

2)

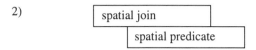

"Find all pairs of geo-objects that intersect themselves".

SELECT A.name, B.name
    FROM geo-objects A, geo-objects B
        WHERE (**intersect** A.geometry B.geometry)

3)

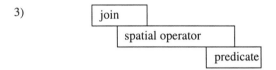

"Compute the intersection of the geometries of the Erft Block with the Venloe Block".

SELECT (**intersection** A.geometry B.geometry)
    FROM geo-objects A, geo-objects B
        WHERE A.name = "Erft Block"
            AND B. name = "Venloe Block"

4)

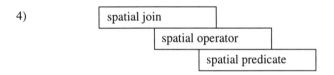

"Compute the intersection of the geometries of the geo-objects *A* and *B*, if they intersect themselves".

SELECT (**intersection** A.geometry B.geometry)
    FROM geo-objects A, geo-objects B
        WHERE (**intersect** A.geometry B.geometry)

With the basic and the advanced building blocks we have introduced a framework for the formulation of spatial queries of a GIS. Interesting is that for the well known set operators[1] of the relational algebra between relations, corresponding operators for the union, the intersection and the difference between geometries can be defined on the level of the spatial operators. The terms *global* and *local* queries, introduced at the beginning of chapter 4.2, can now be grasped more precisely. Global queries always contain advanced building blocks, i.e. they execute a spatial selection, a spatial projection or a spatial join or a combination of these

---

1. Without loss of generality, we have introduced the spatial join instead of the cartesian product as a building block of the relational algebra.

building blocks on several geo-objects, respectively. The upper examples 1 to 4 are global queries. Local queries are restricted to a single geo-object, i.e. they primarily consist of basic building blocks. On the internal sub-structures of a geo-object, however, basic building blocks as well as advanced building blocks can be applied. For example, the distance of two triangles within a triangle network of a geologically defined surface or the volume of a part of a geological stratum can be computed.

4.3 A General Notion of Space

A model for the integration of spatial information for GISs should support all types of spatial operations. We now introduce a general, three-level notion of space for the integration of spatial information. We give some basic definitions that are taken from standard literature (QUERENBURG 1979; DTV 1991). Depending on the degree of abstraction we later intend to find out on which "level of the notion of space" a GIS-query can be answered.

Level 1: The topological space

For a better understanding of the topological space we first introduce some basic topological terms. We refer to QUERENBURG (1979) and DTV (1991).

Considering the $\Re^2$ $[\Re^3]$ with open disks[1] [spheres], i.e. disks [spheres] without boundary, buffers around a point P can be described. The open disks [spheres] around P with arbitrary radius and each of their upper sets are called *environments of P* in $\Re^2$ $[\Re^3]$.

Let M be a point set. Then we distinguish between *touching points* and *exterior points* of M. The first are such points for which each environment has at least one point in common with M. The points of M for which this is not true are the exterior points. Touching points can be subdivided into *interior points,* i.e. a complete environment belongs to M for every point and *boundary points,* i.e. no complete environment belongs to M. The set of all interior points of M we call *interior of M* and the set of all boundary points of M is the *boundary of M.* An open set is a set that only contains interior points.

The topological space can be defined over *environment axioms* or over open sets. Because of the importance of closed and *open sets* in the following context, we here introduce the second definition.

Definition 4.1: (M; μ) is called *topological space*[2], if μ is a subset from π(M) with the following properties[3]:

(O1) $\emptyset \in \mu, M \in \mu,$

(O2) $O_1, O_2 \in \mu \Rightarrow O_1 \cap O_2 \in \mu,$

(O3) $\rho \subseteq \mu \Rightarrow \underset{O \in \rho}{\cup} O \in \mu.$

---

1. An open disk is an example for an open set that will be introduced below.
2. "(M; μ)" is to be read as "M provided with the topology μ".
3. Instead of the usually used Gothic letters we here use Greek letters. π(M) means the potential set of M.

μ is called *topology* on the *carrier set M*. The elements of μ are called *open sets,* those of M are called *points*.

Level 2: Metrical spaces

An important class of topological spaces are the so called metrical spaces.

Definition 4.2: (M; d) is called *metrical space,* if a mapping d: M x M $\rightarrow$ $\mathfrak{R}_0^+$ (so called metrics on M) with the following properties exists $\forall$ x, y, z $\in$ M:

(1)	$d(x, y) = 0 \Leftrightarrow x = y$		(identity axiom),
(2)	$d(x, y) = d(y, x)$		(symmetry axiom),
(3)	$d(x, z) \le d(x, y) + d(y, z)$		(triangle inequation).

d(x, y) is called *distance* between *x* and *y*, which refers to a metrics on *M*. On *M*, different metrics can be defined (see chapter 4.7.4). Distances between arbitrary geo-objects, for instance, can be defined on characteristic points like the centre of gravity or minimal circumscribing cubes, i.e. boundary boxes. As we will explain more detailed in chapter 4.7.4, for arbitrary geo-objects not all of the three axioms are valid.

Level 3: The Euclidean space

On the set $\mathfrak{R}^n$ of the n-tuple of real numbers x = $(x_1, ..., x_n)$, y = $(y_1, ..., y_n)$ a metrics is defined by

$$d(x, y) = \|x - y\| = \sqrt{\Sigma (x_i - y_i)^2}$$

The $\mathfrak{R}^n$ provided with this metrics is called *n-dimensional Euclidean Space* and *d* is called *Euclidean metrics.*

Absolute and relative spatial reference

The topological space intuitively implicates a "relative spatial reference", i.e. topological queries refer to the relative position of geo-objects. The same is true for metrical spaces, as metrical queries are also not refered to an absolute coordinate system. In contrary to that, the Euclidean space implicates an "absolute spatial reference", as the Euclidean coordinate system and thus geometric queries directly refer to the coordinates of the geo-objects.

Spatial abstraction

We are now ready to assign the different types of spatial operations to the three levels of the general notion of space. On the first level the *topological operation*s like intersection and inclusion have their place. On the second level *metrical operation*s like distances are placed and finally on the third level the *direction operations* like *aboveOf, belowOf* are standing. Considering the spatial representations of the geo-objects, a "spatial abstraction" can be localized: Euclidean queries demand the absolute coordinates of the geo-objects. With metrical queries, like the distance query between bounding boxes of geo-objects, an abstraction from the coordinates takes place. Not all the coordinates, but only "abstract geometries", i.e. the bounding boxes, are subject of the query. Not the single coordinates of the geo-objects are of interest in the result of metrical queries, but only the distances between the abstract geometries. Finally, topological queries demand neither coordinates nor a metrics, but more general concepts like *"boundary"* or *"interior"* of open sets. Therefore an abstraction can be identified at the level of the spatial representations and at the level of the spatial operations.

Fig. 4.1 shows an example of a "spatial abstraction". Let two closed polygons be given in a plane of the Euclidean space by their points $P_1, ... P_6$, and $P_1', ... P_6'$ respectively, with $P_i = (x_i, y_i, z_i)$ and $P_i' = (x_i', y_i', z_i')$, i = 1, ..., 6. Let their distance be defined by the distance between the centres of their circumscribing rectangles. In the topological space the position of the both polygons is represented by means of two disjoint open sets[1].

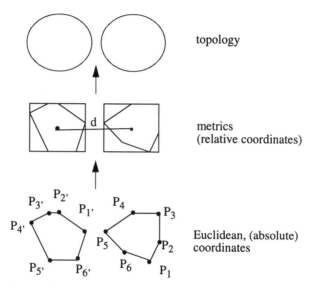

Fig. 4.1 Example of a spatial abstraction

The geometry, i.e. the coordinates of the geo-objects, does not play any role. Let us discuss three variants for the *integration of the three levels* of the general notion of space. Our goal is a unified support of spatial queries, which is independent of the type of spatial operations and spatial representations.

---

1. BERKEL et al. (1988) use versions to represent the different abstraction levels of complex CAD-objects from the design to the detailed drawing. However, the direction of abstraction is exactly in opposite to the just shown example: from the abstract design object to a special CAD-object.

Integration of the three levels

Variant 1:

The spaces are treated separately, i.e. the topological queries are exclusively determined by the topology of the geo-objects. Metrical queries are detemined by the metrics, i.e. relative coordinates, and finally "Euclidean", i.e. geometric queries, are determined by means of absolute Euclidean point coordinates (x, y, z).

Variant 2:

The Euclidean space is a special case of a metrical space. Thus the metrical spaces can be considered as special topological spaces. As a result topological, metrical and Euclidean queries are computed by the comparison of absolute coordinates.

Example:
Let us consider the intersection, the distance between two geo-objects and the access to the coordinates of a set of geo-objects (see fig.4.2a) in both variants. Additionally, let us examine the neighbourhood of two polygons in an area partition, the distance of two polygons of the area partition and the access on a subset of the coordinates (window-query) of the area partition (see fig. 4.2b).

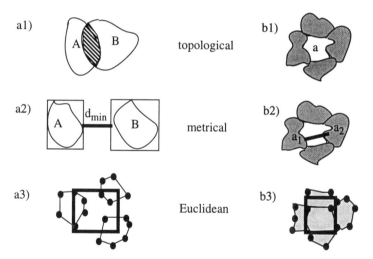

Fig. 4.2 Examples of global (a) and local (b) queries

Variant 1:
To determine the intersection of the geo-objects A and B (fig. 4.2 a1) topologically, i.e. without a metrics and without an access to coordinates, the intersections between the boundary @ and the interior ° ($[@ \cap @]$, $[@ \cap °]$, $[° \cap @]$, $[° \cap °]$) of A and B must be explicitly precomputed. The geometry of the intersection between A and B cannot be topologically determined. The distance between A and B (fig. 4.2 a2), for instance, can be computed by means of the minimal distance between the bounding boxes of A and B. The set of all geo-

objects that intersect a specified query window (fig. 4.2 a3) can be determined by a comparison of the Euclidean coordinates of the query window and the geometries of the geo-objects.

For the determination of the neighbourhoods of area *a* (fig. 4.2 b1) those areas must be determined that have a common boundary with *a*. The distance between two faces in an area partition (fig. 4.2 b2) and the determination of such faces of the area partition that intersect a query window (fig. 4.2 b3) can be answered in the same way as for the global query.

Variant 2:
The intersection of the geo-objects *A* and *B* (fig. 4.2 a1) is determined by a comparison of all coordinates of *A* and *B*. Thus the geometry of the intersection between *A* and *B* can be determined. The same is true for the computation of the distance between *A* and *B* (fig. 4.2 a2) and for the window query (fig. 4.2 a3).

The neighbours of *a* in the area partition are determined by a comparison of the coordinates of all faces of the area partition (fig. 4.2 b1). Neighbours of *a* are those faces that have common points with face *a*. The distance between two faces in the area partition (fig. 4.2 b2) and the window query (fig. 4.2 b3) are to be answered analogously to the global case.

Discussion of the variants:

A disadvantage of variant 1 is that the intersections between the boundary and the interior of all geo-objects must be computed a priori for global queries. Furthermore, in this variant we loose precision during distance computations, as they are not based on the exact geometry of the geo-objects. A disadvantage of variant 2 is that for local topological queries like neighbourhoods, unnecessary many comparisons on coordinates, for example for all faces of an area partition, have to be executed. To eliminate the disadvantages of both methods, a *hybrid approach* can be taken into consideration that combines variant 1 and variant 2.

Variant 3:
We propose to use variant 1 for queries on a single geo-object (local query) and variant 2 for queries on several geo-objects (global query). Thus *local topological relationships* like the determination of the neighbourhood of a triangle in a triangle network are answered by means of the explicitly stored neighbourhoods. *Global topological relationships* like the relationships between different triangle networks, are not explicitly computed, because the costs would be to high with large data sets. As we will see in chapter 4.5, the local topology of the corresponding geo-objects can also be exploited for the increase of efficiency of geometric algorithms.

With the classification of spatial operations and queries and their ordering within a general notion of space, we have marked out the framework for an integration of spatial information for GIS. In the following we intend to turn ourselves towards the data structures and operations of a model for the integration of spatial information. Our goal is to develop a *unified spatial representation for geo-objects* that considers all of the three levels of the introduced notion of space, i.e. geometry, metrics and topology of the geo-objects.

4.4 Extended Simplicial Complexes

In chapter 4.2 we posed the question, if the differentiation into queries on substructures of a single geo-object (local queries) and into queries on several geo-objects (global queries) should also influence the choice of the spatial representation of the geo-objects. For global queries, the geometries of the geo-objects can be considered as "values" which can be "calculated". During the intersection of a geometry of a geo-object with another geometry, for instance, the intersecting geometry is computed. In principle, this is nothing different than, for example, the addition of two integer numbers which generates, as well known, a new integer number. In local queries, however, the geometry of a geo-object is "more" than only a value. The geometry consists of sub-geometries (for instance triangles of a triangle network), i.e. it consists of many "values". Additionally, the internal topology, i.e. the neighbourhoods of the subgeometries must be considered from the point of view of a user. As a result of the discussion of the three variants for the integration of the three levels of the general notion of space (chapter 4.3), for *global* queries a spatial representation should be chosen that exclusively describes the geometry. Separated from that for *local* queries a representation should be taken that contains sub-geometries and the topology (see also fig. 4.2). In this way topological and geometric queries would be explicitly supported. However, we will see in chapter 4.5 that the internal topology can also be used for global queries to increase the efficiency of geometric algorithms. Furthermore, the specification of global topological relationships can be achieved in an elegant way by using the topology of the geo-objects. Thus we strive for a *unified spatial representation* for 3D-GISs, independently of the type of the spatial query (i.e. local and global query, respectively).

In chapter 2 we introduced the best known spatial representations. However, these representations considered topology not at all or only insufficiently. Furthermore, for most of the representations the realization of efficient geometric operations is an open problem. Mostly intersection and inclusion operations on arbitrary polygons and polyhedra are very costly[1]. However, often only very small parts of the complete geometry are relevant for spatial queries (BRINKHOFF et al. 1993). Thus it is nearby to decompose the objects into sub-objects like triangles and executing the operations only on those sub-objects. The intersection of triangles can be reduced to simple line- and plane-intersections, respectively. The subdivision of an arbitrary polygon into triangles is called *triangulation*.

We can motivate the use of simplicial complexes as a basis of a spatial representation for 3D-GISs in several ways. First, it is possible to represent points, lines, surfaces and volumes in a unified model. Second, only "well defined" surfaces can be build by simplices of the dimension 2 (triangles), as a triangle cannot intersect itself[2]. A decomposition of volumes into simplices of the dimension 3 (tetrahedra structures) is favourable, because tetrahedra are always convex and because they have a small and constant number of faces, edges and points. The consequence is that only few cases have to be treated separately in the geometric algorithms. Furthermore, the faces of a tetrahedron consist of triangles which are always convex. Thus a decomposition of volumes into simplices of the dimension 3 is defined in that way

---

1. Mostly $O(n^2)$ with regard to the number n of the polygons.
2. This is no more valid for the 4-corner, as it can be seen with the well-known Moebius-strip, which arises, if we turn a rectangle and compose the ends in that way that the diagonal corners fit together.

that it consists for each dimension of the most simple geometry, respectively, i.e. point, line, triangle and tetrahedron. Third, in the geoscientific practice often irregular distributed measure points are caught.Their representation as regular grid leads to big inaccuracies, whereas the measure points can be suitably interpolated on the surface as triangle irregular networks (TINs). Because of the higher precision, this representation is also particularly suited for the interactive modeling of geological defined strata models (SIEHL et al. 1992).

We extend the simplicial complex to use it as the kernel of a model for the integration of spatial information. The extension we call *"e-complex" (extended complex)*. As it is well known, a 0-simplex is a vertex, a 1-simplex an edge, a 2-simplex a triangle and a 3-simplex a tetrahedron. We extend a d-simplex in the following way:

Definition 4.3: A *d-simplex with neighbourhood (d > 1)* consists of a d-simplex and of maximal *(d + 1)* adjacent simplices of dimension *d* ("neighbouring simplices").

Fig. 4.3 shows a 2-simplex with three 1-simplices (the edges *v1 - v3, v3 - v2* and *v2 - v1*) and three adjacent 2-simplices (the triangles *t1, t2* and   ) as the neighbourhood. The adjacent simplices can also be "empty", i.e. the neighbours need not exist.

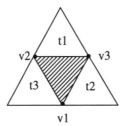

Fig. 4.3 d-simplex with neighbourhood for d=2

A "2-simplex with neighbourhood" can be described by the following 7-tuple[1]:

> *2-simplex (notation, vertex1, vertex2, vertex3,*
> *neighbouring triangle1, neighbouring triangle2, neighbouring triangle3).*

The *notation* gives the name of the geo-object, whose topology is described by the simplex. To avoid redundancy, we renounce the explicit description of edges. The three vertices and the three neighbouring triangles belong to a simplex, respectively. In the following the explicit storing of neighbouring triangles will take an important role, especially if we intend to access to single simplices of a complex. Correspondingly, four vertices and the four neighbouring tetrahedra belong to the topology of a 3-simplex.

Our goal is to develop a spatial representation that bases on the simplicial complex and explicitly considers the topology and geometry of geo-objects. Thus a separated access to the topology and the geometry of the geo-objects is to be enabled.

---

1. This representation is also particularly suited for a compact storage in a database (see chapter 6).

Definition 4.4: An *e-complex* is a triple *(C, T, G)*. *C* is a simplicial complex of dimension *d*. *T* is the set of the d-simplices (for *d > 1*: with neighbourhood) of *C*. *T* is called the d-dimensional topology of *C*. *G* is the Euclidean geometry of *C*, i.e. the set of the Euclidean coordinates of the 0-simplices of *C*.

Remark:
a) Let *(C, T, G)* be an e_0-complex, then: *T* is the set of vertices (0-simplices) of *C*.

b) Let *(C, T, G)* be an e_1-complex, then: *T* is the set of connected, eventually forked edge sequence of *C* (see fig. 4.4) that consists of a list of the 1-simplices (edges) that again are composed by two adjacent 0-simplices (vertices).

c) Let *(C, T, G)* be an e_2-complex, then: *T* is the set of triangle networks of *C*, which consists of a list of 2-simplices (triangles) with neighbourhood that again are composed by three adjacent 1-simplices (edges). Every 1-simplex again consists of two adjacent 0-simplices (vertices).

d) Let *(C, T, G)* be an e_3-complex, then: *T* is the set of tetrahedron networks of *C*, which consists of a list of 3-simplices (tetrahedra) with neighbourhood that again are composed by four adjacent 2-simplices (triangles). Every 2-simplex again consists of three 1-simplices and each 1-simplex is composed by two 0-simplices.

Fig. 4.4 Forked line sequence of an e_1-complex

Remark:
In *T* "bays", "holes", "rings" and "islands" are allowed. Let at least one metrics on *G* be defined. *G* is the set of the (x, y, z)-coordinates of the 0-simplices of the e-complex. We say that an e-complex is of dimension *d*, if the highest dimension of its simplices is *d*. We call the e-complex of dimension d "e_d-complex".

Prior to the introduction of the terms "bay", "hole", "ring" and "island" (see fig. 4.5) we define the boundary and the interior of an e_d-complex.

Definition 4.5:
a) *The boundary* @ *of an* e_d*-complex A (d > 0)* is the set of its (d-1)-simplices for which is valid: In each of their environment is as well a point which belongs to the e_d-complex as a point which does not belong to the e_d-complex.

b) The *exterior boundary* @$_{ext}$ of an e_d-complex A *(d > 0)* is the boundary for which is valid: There is no topology of A that is outside @$_{ext}$.

c) An *interior boundary* @$_{int}$ of an e_d-complex A *(d > 0)* is a boundary for which there is at least one topology of A that lies outside @$_{int}$.

Definition 4.6: The interior of an e_d-complex *(d > 0)* is the topology of the e_d-complex without its exterior boundary and its interior boundaries.

In the following we informally introduce the terms *bay, hole, ring* and *island* of an e_d-complex (for $2 \leq d \leq 3$). A *bay of an e_d-complex* is a topology that lies between its exterior boundary and its convex hull (see fig. 4.5). We call a topology a *hole of an e_d-complex*, if it has an interior boundary of the e_d-complex as an exterior boundary. A *ring of an e_d-complex* is a topology with exactly one hole. Finally an *island of an e_d-complex* is a topology that has an interior boundary of a hole as exterior boundary and which itself is no ring.

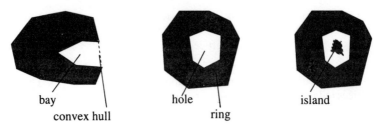

bay   hole   island

         convex hull   ring

Fig. 4.5 Bay, hole, ring and island of an e_2-complex

In the following we already point out a speciality of e-complexes without giving a formal description[1]: At one vertex of the e-complex not more than one hole may meet, i.e. no holes lying "opposite to each other" may occur (see fig. 4.6).

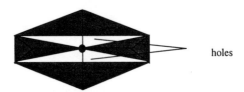

                                                                    holes

Fig. 4.6 Not allowed "opposite lying holes" of an e_3-complex

Let us closer identify the topology for e-complexes of different dimension d $(0 \leq d \leq 3)$. An important aspect for the design of a spatial representation is its interaction with efficient geometric algorithms. Emphasizing this point we introduce the *convex e-complex*. To simplify the computation of spatial operations, we also define the *connected e-complex*.

Definition 4.7: An e-complex of dimension $d > 0$ is called *connected,* if every two of its 0-simplices are connected by a sequence of 1-simplices, respectively.

Definition 4.8: An e-complex A is called *convex e-complex* in 2- and 3-dimensional Euclidean space, respectively, if for each of its connected e-complexes is true: For every two of its 0-simplices, respectively, its connecting line also is inside A.

---

1. This restriction has to be made because of the geometric algorithms which will be introduced in chapter 4.5.

Explanations:

An e_1-complex is called a *convex e_1-complex,* if the 2- and 3-dimensional space, respectively, between the e_1-complex and its convex hull is filled by 2-simplices (triangles in 2D space) and 3-simplices (tetrahedra, in 3D space), respectively.

An e_2-complex is called a *convex e_2-complex,* if the 2- and 3-dimensional space, respectively, between the e_2-complex and its convex hull and its holes are filled by 2-simplices (triangles in 2D space) and 3-simplices (3D space), respectively.

An e_3-complex is called a *convex e_3-complex,* if the 2- and 3-dimensional space between the e_3-complex and its convex hull and its holes are filled by 3-simplices (tetrahedra).

The triangles and tetrahedra, which are generated during the computing of convex e-complexes[1] we call *virtual triangles and tetrahedra,* respectively. The original triangles and tetrahedra are called *real triangles and tetrahedra,* respectively. Fig. 4.7 shows examples for (convex) e-complexes of the different dimensions in 2D- and 3D-space.

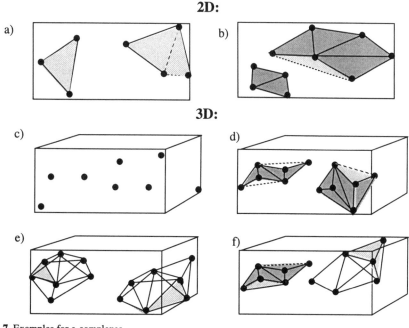

Fig. 4.7 Examples for e-complexes
    a) (convex) e_1-complex in 2D space   b) (convex) e_2-complex in 2D space
    c) e_o-complex in 3D space   d) (convex) e_2-complex in 3D space
    e) (convex) e_3-complex in 3D space   f) "mixed" (convex) e-complex,
                                        composed of 2- and 3-simplices

---

1. The idea to introduce *convex* e-complexes sprang up during a discussion about efficient geometric algorithms with my colleague Hermann Stamm-Wilbrandt.

4.4.1 Generation of Convex e-Complexes

The most obvious motivation to use convex e-complexes as a spatial representation for geo-objects is the essential simplification of geometric algorithms using this representation. In chapter 4.5 we give efficient algorithms on convex e-complexes. Let us now introduce a method for the generation of convex e_2-complexes in 2D and 3D space as well as for convex e_3-complexes in 3D space. Therefore we first introduce some terms to be used for the following algorithms.

A *local bay* of an e_d-complex *(d > 1)* is a bay for which is valid: The angle between neighbouring boundary simplices of the dimension *d* does not change between an acute and an obtuse angle. A *global bay* is a bay that is still existing after the filling up of all local bays. A global bay again may contain one or several local bays.

A *local hole* of an e_d-complex *(d > 1)* is a hole for which the angle between the neighbouring boundary simplices of dimension *d* does not change between an acute and an obtuse angle.

A *global hole* of an e_d-complex *(d > 1)* is a hole that still exists after the filling of all local holes. A global hole again may contain one or several local holes.

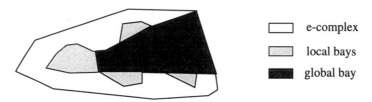

☐	e-complex
▨	local bays
■	global bay

Fig. 4.8 Local bays and global bay

Fig. 4.8 shows an example for local bays and a global bay which still exists after the filling up of all local bays. We now introduce a simple method for the filling up of bays and holes in 2D space.

algorithm *MakeConvex2D (A)*

{Input is a concave e_2-complex *A* in 2D space. Fill *A* so that it becomes a convex e_2-complex}

For the exterior boundary [and all interior boundaries, respectively]:
  While the considered boundary has at least one bay [one hole, respectively]:

Step 1:   Start at an arbitrary edge of the boundary for which the outside angle to the next edge is larger [smaller, respectively] than 180 degrees, i.e. the angle is obtuse [acute, respectively];

Step 2:   Go to the next boundary edge[1], respectively, until the outside angle between the current and the previous edge is smaller than 180 degrees or the first edge is reached again. Call the previous edge *start edge;*

---

1. For example anti-clockwise.

Step 3:   While the outside angle is smaller [bigger, respectively] than 180 degrees and
              the first edge is not yet reached:
              Connect the end point of the current boundary edge with the start point of the
              start edge, i.e. fill the bay and the hole, respectively, with a triangle. Go to the
              next boundary edge;

Step 4:   If the first edge is not yet reached, goto step 2;

Step 5:   Insert the new neighbours for the new generated, virtual triangles and their
              neighbouring triangles into *A*

end *MakeConvex2D.*

MakeConvex2D fills all (local and global) bays at the exterior boundary of the e_2-complex
and the (local and global) holes at all interior boundaries. Fig. 4.9 a) and b) show the filling
up of a local bay and a local hole of an e_2-complex, respectively. For simplification, in fig.
4.9 a) only the boundary of the e_2-complex and in fig. 4.9 b) only the boundary of the hole
of the e_2-complex is drawn[1].

Complexity of MakeConvex2D:

Obviously the filling up of the bays and holes (steps 1 - 4) is possible in $O(n + m)$, let n be
the number of boundary edges of the e-complex and let m be the number of boundary edges
that are new generated. Thus the *steps 1 - 4* have a linear running time refering to the number
of new generated "virtual" triangles. That is why the following Euler polyhedron formula
holds for graphs, which are composed by triangles[2]:

$$\#triangles = \#boundary\ edges - \#vertices + 2.$$

Step 5 also needs linear time refering to m. The above algorithm for the filling up of bays and
holes has the disadvantage that it can lead to "acute" triangles. "Better" fillings can be
achieved, if in a preliminary step, the convex hull of the e_2-complex is computed by methods
like the Graham-Scan (PREPARATA & SHAMOS 1985).

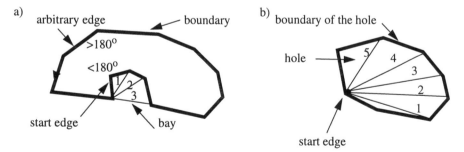

Fig. 4.9 Filling up of a local bay (a) and of a local hole (b) for e_2-complexes in 2D space

---

1. We have numbered the edges that are generated during the filling up by 1 - 5.
2. "#" stands for "number of".

The algorithm *MakeConvex2D* can be directly transfered to the threedimensional case, i.e. the filling up of the e_3-complexes to convex e_3-complexes. Then in the *steps 1 - 3* the angle between the boundary triangles has to be considered instead of the angle between boundary edges. A more serious problem is the filling up of e_2-complexes (triangle networks) in 3D space. Hitherto, algorithms which construct the convex hull from a point set in threedimensional space, either have a quadratic running time or are based on the data structure of a double referenced list (so called gift-wrapping techniques, see PREPARATA & SHAMOS 1985). We introduce a method that in its first part, i.e. for the computation of the convex hull, is based on the *quickhull*-algorithmus of BARBER et. al. (1993).

Let us first introduce some terms for a better understanding of the algorithm . Let d be the dimension of a Euclidean space and let P be a point (0-simplex) of an e-complex.

Definition 4.9: A *facet F* is the *(d - 1)*-dimensional boundary of a hull.

Comment: Let an order be defined on the points of each facet so that "above" is defined in any location outside the hull.

Definition 4.10: A *visible facet VF* of a point P is a facet for which P is above this facet.

Definition 4.11: A *horizon* of a point P is the exterior border of all visible facets for P.

Definition 4.12: A *ridge* is the *(d - 2)*-dimensional border of a facet.

algorithm *Quickhull (A)*

{Input are the points (0-simplices) of an e_d-complex A. Determine the convex hull of A}

Step 1:   Generate a convex set of (d+1) linear independent points (0-simplices) and call them *initial hull of A;*

Step 2:   Partition the remaining points over the *"outside-sets"* (point sets that are lying above the visible facets) of the initial hull:

>   Find for every point a visible facet;
>   *If* there is none, *then* the point is already an interior point and need no more to be considered;
>   *else* determine the distance between the point and the centre of the visible facet;
>
>   Insert the point into the outside set of this facet;
>   Store for each facet the point which has the largest distance to this facet;

While facets are existing whose outside set is not empty:

>   *Step 3:*   Take the point which has the largest distance from the initial hull as new point.[1]

>   *Step 4:*   Determine for this point the horizon and the corresponding visible facets:

---

1. If there are several points which have the same distance, then take the first you find.

Iteratively determine for all neighbours of the visible facets, if they are visible[1]. The exterior ridge of these facets forms the horizon;

Step 5:   Generate a "cone" of new facets from the point to the horizon, i.e. build new facets that have the ridges of the horizon as basis and the new point as peak;

Step 6:   Partition all other points of the facets that are inside the cone over the new facets. Delete the visible facets inside the horizon

Step 7:   Return the convex hull of *A*

end *Quickhull.*

Comment: In step 2 the user may give the option that for every point the algorithm does not already terminate after the first found visible facet, but not till all visible facets are searched through. Then the point is inserted into the outside set of the facet which has the largest distance to the facet.

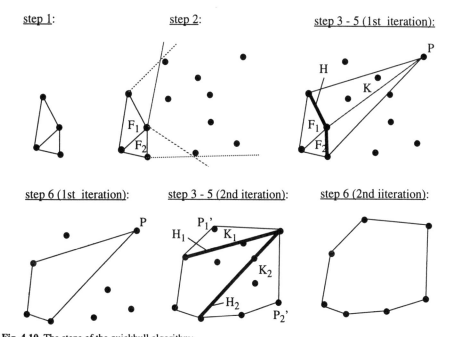

Fig. 4.10 The steps of the quickhull algorithm:
Initial hull *(step1)*,
Partition the remaining points into outside-sets of the visible facets [here: F_1, F_2] (step 2),
Determination of the point with the largest distance from F_1 and F_2, of the *horizon H* of point *P* and of the cone *(step 3 - 5)*,
Deletion of the visible facets inside the horizon *(step 6)*

Fig. 4.10 shows an example for the different steps of the quickhull algorithm. In the following we introduce the complete method for the filling up of e_2-complexes in 3D space[2]. The method consists of the two following parts:

---

1. This is straight forward, as every facet knows its neighbours.

1) Determination of the convex hull of the points (0-simplices) of the e_2-complex,

2) Generation of tetrahedra between the triangles (2-simplices) of the e_2-complex and convex hull (see fig. 4.11).

Figuratively spoken, from a triangle network in 3D space, a convex tetrahedron network is generated by the filling up with tetrahedra. Let a *boundary triangle* of an e_2-complex be a triangle of which at least one edge is lying on the boundary of the e_2-complex.

algorithm *MakeConvex3D (A)*

{Input is an e_d-complex A in 3D space ($2 \leq d \leq 3$). Fill up A to a convex e_d-complex}

Step 1:   *Quickhull (A);*

Repeat until the first boundary triangle is reached again:

Step 2:   Search a boundary triangle of A with the property that one of the points of the boundary triangle is a point of the convex hull. Let this point be P_1;

Step 3:   *While* a second point of the boundary triangle is also a point of the convex hull: Examine the next (neighboured) boundary triangle;

Step 4:   Examine the next boundary triangles *until* a boundary point of A is again a point of the convex hull. Let this point be P_2. Store all the visited triangles;

Step 5:   Adapt the boundary of A to the convex hull (see figure 4); generate tetrahedra: Connect every boundary point of A between P_1 and P_2 with P_1 (or with P_2, i.e. generate tetrahedra between A and the convex hull; insert "virtual" triangles between the boundary and the convex hull of A, so that every edge of the new boundary becomes a ridge of the convex hull. Goto step 2;

Step 6:   Fill up the holes (on the triangle network):
Repeat until all boundary triangles are run through:
Mark every examined boundary triangle;
Triangulate the holes, with the well known Delaunay method and generate the tetrahedra between the virtual triangles and the convex hull of A;
Return A

end *MakeConvex3D.*

Step 1 determines the convex hull of A. In the *steps 2 to 5* the bays of the e_d-complex A are filled up and finally in *step 6* the holes are closed.

---

2. The method was implemented on top of the ODBMS ONTOS (1992) (see STIENECKE 1995) and is to be integrated into GEOSTORE (BODE et al. 1994).

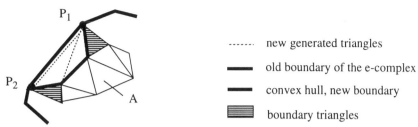

Fig. 4.11 Adapting of the boundary of an e_2-complex A to the convex hull (sketch in 2D space)

In the following we show the correctness of *MakeConvex3D*. The first part *(quickhull)* picks out the essential parts of the proof by (BARBER et al. 1993). However, it is more closely orientated to the implementation of the algorithm. The second part shows the correctness of the tetrahedron filling.

Correctness of MakeConvex3D:

a) Determination of the convex hull (quickhull):

Let d be the dimension of a simplex. By the definition of the d-simplex the initial hull that is generated by $(d + 1)$ linear independent points, is a convex hull *(step 1)*. Thus it is to show that the remainder of the method generates a convex hull of the given point set in any case. Thus we first have to prove that every extreme point is processed in the algorithm that lies above two or more facets during the partitioning step of *quickhull (step 2)*.

Proof a1): Assume the opposite, i.e. let P be an extreme point that lies above at least two facets and that is not processed in the *steps 3 - 6*. Let P be allocated to a facet F, i.e. let P be inside the outside-set of F. By assumption there must be a point which has the largest distance whose "cone" is above or coplanar with P and whose visible facets enclose F. Then P is, however, an interior point of the hull. Thus it is no extreme point. ♦

Finally we have to show that only extreme points become points of the convex hull during the partitioning into outside-sets *(step 2)* and during the processing of the point with the largest distance from a facet *(steps 3 - 6)*.

Proof a2): Assume the opposite, i.e. let P be a point that becomes a point of the convex hull by the partitioning and which is no extreme point. Let G be the convex hull *conv($p \cup H$)*. After the localization of a visible facet *(step 4)* a cone of new facets is generated. Thus, if P is inside H, then P cannot become a point of the convex hull *conv($p \cup H$)*. If P is outside H, then a ridge r of H exists, which has a neighbour facet above (i.e. outside H) and a neighbour facet below (i.e. inside H). Hereby the simplex with basis r and peak p is a facet of G. By the proof a1) is shown that every extreme point is processed in the *steps 3 - 6*. By the partitioning in *step 2*, however, P is a point with the largest distance to the facet which is below r. Thus P is an extreme point. ♦

b) Filling up with tetrahedra:

We show the correctness of the remainder of the method *(steps 2 - 7)* as follows: We have to prove that there can be no connecting line between two points inside the convex hull, which is not completely running through the tetrahedra of the e-complex.

Proof: Assume the opposite. Then a point inside the convex hull must exist which is neither inside nor on the boundary of a tetrahedron. As every interior tetrahedron, however, has four neighbours by construction , such a point can neither exist between interior nor between exterior tetrahedra ("there is no place"). Thus the point had to be outside an exterior tetrahedron, but inside the convex hull. However, this cannot be true, because the exterior tetrahedra are exactly the facets of the convex hull by construction. ♦

Let us estimate the complexity of *MakeConvex3D*.

Complexity of MakeConvex3D:

a) Complexity of quickhull:

The complexity of *quickhull* is an open problem. Under certain balance conditions it was shown that the complexity of *quickhull* for *n* input points and *r* output points for $d \leq 3$ is $O(n \log r)$ (BARBER et al. 1993).

b) Complexity for the filling of tetrahedra:

The adaptation of the boundary of the e_2-complex to the convex hull and the tetrahedralisation can be computed in $O(t)$. Let *t* be the number of triangles. Then $O(t)$ is to be estimated for $(d \leq 3)$ by $O(n)$ as an upper border.

Differentiation of the interior boundaries and the exterior boundary:

During the search for the boundary of the object one can also meet an "interior boundary" (hole). Thus we still have to exclude this case (see fig. 4.12).

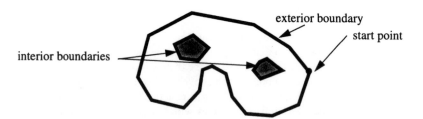

Fig. 4.12 Differentiation of interior boundaries and exterior boundary

algorithm *TestBoundary (@A)*

{Input is a boundary @A of an e_d-complex *(2 ≤ d ≤ 3)*. Determine if @A is an exterior or an interior boundary}

Step 1:   Mark a point from which is known that it is an exterior boundary point (start point);

Step 2:   Run through the boundary completely. If you meet the start point again, the boundary is the exterior boundary of the e_d-complex, otherwise it is an interior boundary)

end *TestBoundary.*

4.4.2 Compressed Representation

We can also regard a simplicial 2- and 3-complex as a graph, respectively, because their triangulations consist of 0- and 1-dimensional simplices. Thus it is possible to use *compression techniques for graphs* to reduce the data volume of e-complexes. We distinguish two cases:

Case 1:

The needed storage place is no problem for the database, but for main memory. Then, one should be able to load sub-complexes into main memory like "all parts of an e-complex which are inside a given box". Hereby spatial access methods for e-complexes may help. They will be closer examined in chapter 5.3.

Case 2:

The needed storage place is also a problem for the database. Then it is convenient to store the e-complexes compressed in the database and to load sub-complexes, when required, decompressed into main memory. Thus the compressed representation can also be used for the data transfer between different GISs.

TURAN (1982) gives a compression method that generates a compressed dual coding from an undirected planar graph by a *depth-first-search* with a minimal spanning tree[1] in *linear time*. The dual coding reduces the data volume for the representation of the graph by a constant factor of 16, considering a storage size of 4 byte per vertex of the graph.

---

1. A minimal spanning tree is an undirected planar graph *G* with the complete number of vertices for *G*, but a minimal number of edges.

4.5 Transformation of Spatial Representations into e-Complexes

To execute geometric operations on objects of different spatial representations (let for example an octree-object intersect an e-complex object), we have to execute a transformation into the e-complex representation. However, this may be avoided if the bounding-box test was positive before. Thus the bounding-box has a representation-overshooting function.

In the following we consider transformations of spatial representations into e-complexes, first in 2D space and afterwards in 3D space.

4.5.1 Transformations in 2D Space

Taking up the notation from chapter 2.4, let the mappings

$$ov :\ O \to V, \quad or :\ O \to R, \quad oe :\ O \to E$$

be given, which map objects of the object space O into objects of the vector space V, the raster space R and into the e-complex space E. Furthermore, let the transformations

$$(ve :\ V \to E\ ;\ ev :\ E \to V) \quad \text{and} \quad (re :\ R \to E\ ;\ er :\ E \to R)$$

be given, which transfer the objects of the vector space V and the raster space R into objects of the e-complex space E and vice versa (see fig. 4.13).
Then:

(An $o \in O$ may exist | re (or(o)) $\neq$ oe (o)) $\wedge$ ($\forall\, o \in O$: ve (ov (o)) = oe (o))

Proof:

Part (1):   Assumption: $\forall\, o \in O$: re (or(o)) = oe(o).

This was already shown in the proof of theorem 2.1 in chapter 2.4, as a simple vector is also the simplest case of an e_1-complex.

Part (2):   For e_0- and e_1-complexes is true: $\forall\, o \in O$: oe (o) = ov (o), i.e.. $V = E$
$\Rightarrow \forall\, o \in O$: ve (ov (o)) = oe (o)

For e_2- and e_3-complexes[1]:

Assumption: $\exists\, o \in O$: ve(ov (o)) $\neq$ oe (o)

$\Rightarrow$ as *ve* does not change the boundary of *ov*(o) and
@(ov(o)) = @(oe(o)):

$\Rightarrow$ *ov*(o) changes @o or the interior ^{o} of *o* (contradiction). $\blacklozenge$

---

1. In the following @(o) stands for "boundary of o".

We show that the transformation of an object along the raster space leads to a loss of information, which is drawn hatched in fig. 4.13. In the transformation of an object along the vector space, we receive the same result as in the direct transformation of the objects into the e-
complex, if we consider only the boundary @o of the objects.

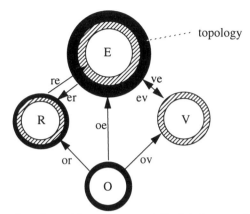

Fig. 4.13 Transformation from the raster- and the vector representation into the e-complex

During the transformation from the object space O into the vector space V, precision is lost
because only a more or less precise approximation of the geometry can be digitally represented. During the mapping of O into the raster space R also information is lost by the approximation which is dependent on the resolution of the raster. During the transformation
from V and R into the e-complex space E additionally an *internal topology* is added by the
triangulation (see the grey ring in fig. 4.13). It may be chosen depending on the original topology in object space (see fig. 4.14), if the e-complex is directly generated from the object
space. This internal topology, however, is lost during the back transformation.

An example:

Let us consider a geo-object "Berlin-map", which shall be transformed directly and along
the detour of the raster and vector representation, respectively, into the e-complex representation (see fig. 4.14).

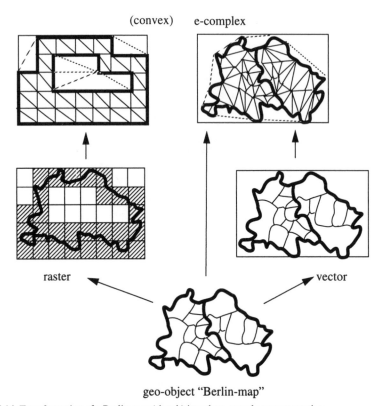

geo-object "Berlin-map"

Fig. 4.14 Transformation of a Berlin map (sketch) into the e-complex representation

The e-complex representation of Berlin, which is derived by the raster representation, also is an imprecise approximation, because the approximation of the raster representation[1] already is imprecise. During the trivial triangulation of the raster cells the topology (neighbourhoods of the raster cells) is preserved. In the e-complex representation, which originated by a transformation from the vector-representation, the boundary for not-convex e-complexes is unchanged against the vector representation. The topology, however, is lost during the mapping from the object space into the vector space. During the generation of the e-complex from the object space and from the vector representation, a new topology is generated by the triangulation, which depends on the original topology of the Berlin map. The triangulation can take place as described for the generation of convex e-complexes in chapter 4.4.1 .

---

1. The raster size was chosen extra large for demonstration.

Complexity:

Need for storage place:

If we decompose a polygon into triangles, starting from the vector representation, for triangulations, which do not generate new points, the *Euler formula*

triangles = # boundary points of the polygon - 2.

is valid. For optimized triangulations with new generated points, which avoid too acute angles, the need for storage place depends on the number of the triangles that have to be updated, because of the insertion of new points. Thus an upper border of time complexity for the insertion of an additional triangle per generated triangle is (2 * (# boundary points of the polygon - 2)).

For the trivial method, which divides a raster cell into two triangles, the decomposition of raster cells into triangles needs *(2 * n * k)* of storage place. Let n be the number of the raster cells and let k be the storage place needed for one triangle. If, however, we generate a closed line sequence as boundary of a raster, the triangulation is independent of the raster. Thus we can use the same method as for the vector → e-complex transformation, i.e.:

triangles = # boundary points of the boundary polygon of the raster- 2.

Need for time:

For the triangulations of which no new points are generated, the conversion from the vector to the e-complex representation can be enabled in *O(n)*. Let n be the number of the points of the vector geometry. With the insertion of a constant number of triangles per generated triangle, the running time is only increased by a constant factor.

Obviously the conversion from the raster to the e-complex representation can be executed in *O(n)* by the triangulation with the trivial method. Let n be the number of boundary cells of the raster. For the method which is based on the circumscribing polygon, the same time is needed as for the vector conversion.

4.5.2 Transformations in 3D Space

The transformation of the 3D-representations into an e-complex consists of a triangulation[1] of the surface and a following tetrahedralisation of the volume. Let us assume that CSG is represented by an internal boundary representation. Fig. 4.15 shows the transformation of the spatial representations introduced in chapter 2 into the e-complex space *E*. During the transformation from the function-boundary representation[2], CSG, sweep- and the parametrized representation, small information losses occur during the transformation into the e-complex for polygon-like geometries, because of the imprecise approximation of the e-complex. For the vector-boundary representation, the cell decomposition and the enumeration method, no

---

1. Optimization criteria for triangulation of surfaces in 3D space can be found in (KLESPER 1994).
2. Compare the pointed arrow from "B-Rep" to "e-com" in fig. 4.15.

information losses occur during the transformation into the e-complex, because the boundary of the geometry stays unchanged. With the tetrahedralisation, additionally a topology is added. The transformation into the opposite direction involves an information loss (see hatched circles in fig. 4.15). Let us explain the process of the transformation exemplarily for the boundary representation by 3D-geometries of different complexity.

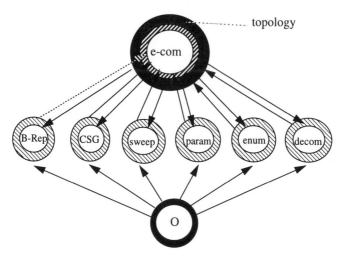

Fig. 4.15 Transformation of the 3D-representations into the e-complex representation

1) Decomposition of the unit cube

We can decompose the unit cube into six tetrahedra[1]. To simplify the decomposition, we decompose the unit cube into two prisms and then decompose the prisms. Thus we reduce the problem (seen from the base) from an n-corner to an (n-1)-corner.

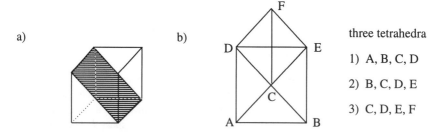

Fig. 4.16 Decomposition of the unit cube into two prisms and tetrahedralisation of a prism

1. There is also a decomposition into five tetrahedra. However this decomposition does not allow a decomposition into two symmetric prisms, which will be described in the following.

In the example the decomposition of a prism (fig. 4.16a) leads to the decomposition of the three tetrahedra with the vertices (A, B, C, D), (B, C, D, E) and (C, D, E, F). Let us apply this simple decomposition of a prism to the decomposition of approximated cylinders into tetrahedra.

2) Decomposition of approximated cylinders

Approximated cylinders (see also chapter 6) have two n-corners as bases. If we tetrahedralize an approximated cylinder, the generalization of the decomposition of a prism leads to

$$(n-2) * 3 = 3n - 6$$

tetrahedra. Fig. 4.17 shows the method for the decomposition. With a "cutting" of prisms we can tetrahedralize them piece by piece as described above. Hereby every prism has to be decomposed into three tetrahedra.

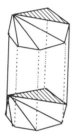

Fig. 4.17 Cutting of prisms for the tetrahedralisation of simple volumes

3) Decomposition of general volumes

For the tetrahedralisation of general volumes (see also LEISTER 1991; KRAAK & VERBREE 1992) a variation of the method for the filling up of convex e-complexes introduced in the *MakeConvex3D*-algorithm (chapter 4.4.1) can be used. As a preliminary step before the tetrahedralisation, the surface of the volumes have to be triangulated, for instance with the well-known Delaunay-method (PREPARATA & SHAMOS 1985). After the triangulation of the boundary the volume has to be filled up with tetrahedra.

4.6 Efficient Geometric Algorithms

In chapter 4.4 we introduced the e-complex as spatial representation of our model for the integration of spatial information. In the following we show that geometric algorithms on e-complexes can be realized efficiently.

The spatial relationships and operators, respectively, which we designated as *inclusion, exclusion* and *intersection*, we rename *inside-, outside- , overlap-relationship* and *overlap-operator*[1], respectively. We give 3D-algorithms on e-complexes for those four operations. They are part of a function library (SCHOENENBORN 1993), realized in the object-oriented database kernel system OMS (BODE et al. 1992), whose algorithms we will evaluate in

chapter 6. For the *overlap-operator* we give different variations (so called *cut-operators*)[1].
We use the so called *bounding-box test*[2] as a filter for each algorithm to reduce the working
set significantly. Hereby the respective geometric operation is executed on the *bounding
boxes,* i.e. on the minimal circumscribing cuboids of the e-complexes, which are directed to
the x-, y- and z-axes. Thus a *preselection* is executed for the simplices which qualify themselves for an operation. For example, for the intersection of two e-complexes *A* and *B* (*over-
lap-operation)* only those simplices have to be taken into account, which intersect the *"intersection box".* These are those simplices, which intersect as well the *bounding box* of *A* as
the *bounding box* of *B*.

The representation for the line and plane equations (1- and 2-simplices) used in the algorithms is the parameter representation, which is given by the following equations:

$$(I) \quad p = p1 + \lambda\,(p2 - p1) \quad and \quad (II) \quad p = p1 + \lambda\,(p2 - p1) + \mu\,(p3 - p1).$$

A point is lying on the line and plane, respectively, iff real numbers λ and μ (with an interval
ε) exist that meet equation I and II, respectively. The *overlap-operator* uses these two equations by processing line-plane intersections, i.e. line- and triangle intersections in the concrete case.

Overlap- and outside-algorithm:

We describe the *overlap-algorithm (overlap-relationship)* only shortly. The test, if two e-
complexes *A* and *B* of dimension *d* intersect each other or not, works as follows: If the bounding boxes of *A* and *B* do not intersect each other, then the result is *false*. Otherwise from the
set of simplices, that intersect as well the bounding box of *A* as the bounding box of *B*, all
d-simplices of *A* are tested for the intersection with all d-simplices of *B*. In the case of e_2-
complexes, for instance, this is an intersection of two triangles, respectively. The algorithm
terminates as soon as a d-simplex of *A* intersects a d-simplex of *B*. Thus it returns the value
true, otherwise *false*.

The *overlap*-algorithm mainly profits from the simplicity of the triangle tests and from the
reduction of the number of relevant geometries by means of the decomposition into triangles.
The latter enables the bounding-box test on simplices. Below we will see that the running
time of the algorithm, which is obviously $O(n^2)$ can be essentially reduced, if the e-complexes are made convex before the processing in the algorithm (see algorithm *overlapOp*). Let *n*
be the number of d-simplices that intersect both bounding boxes.

The *outside*-algorithm works analogously to the *overlap*-algorithm with the difference that
the outside-algorithm terminates as soon as two d-simplices are *not outside* each other. The
overlap-algorithm terminates, if two d-simplices *overlap* each other. Furthermore, this algorithm terminates right at the beginning, if the bounding boxes intersect each other.

---

1. The spatial relationships provide a boolean result, whereas spatial operators again provide a (convex) e-complex as a result.
1. For the implementation of the cut-operators see STIENECKE (1995).
2. See NEWMAN & SPROULL (1981), WIDMAYER (1991).

Complexity of the outside-algorithm:

In the _worst case_[1] of the _outside_-algorithm all triangles of one network have to be tested against all triangles of the other network inside the intersection box. The complexity, however, can be improved by reducing the set of examined triangles and tetrahedra, respectively, for instance with a hierarchical partitioning of bounding boxes like the octree method.

For 2D-objects in 2D space and for 3D-objects in 3D space it is intuitively clear what it means, if two objects are _inside_ or _outside_ to each other. For e_2-complexes in 3D space we define that an e_2-complex A is _inside_ an e_2-complex B, if the surface of A is contained in the surface of B. I.e. for every 2-simplex $A_i \in A$ one or several 2-simplices $B_j \in B$ exist so that A_i is contained in B_j. We can also say that then all triangles of A completely touch the e_2-complex B. Analogously we fix that an e_2-complex A is _outside_ of an e_2-complex B, if none of their triangles touch or intersect each other. Previous _inside_-algorithms which are based on other spatial representations, are mostly limited to the twodimensional case and do not exploit the topology of the geo-objects. In _plane-sweep techniques_ (see PREPARATA & SHAMOS 1985) the complete plane (2D) and the complete space (3D) along a line and a plane, respectively, has to be searched through. We give an _inside_-algorithm for e_2-and e_3-complexes[2], respectively, which exploits the _topology_, i.e. the neighbourhood relationships of the triangles and tetrahedra of the e-complexes, respectively. Thus only those triangles and tetrahedra are searched through that are relevant for the _inside_-relationship.

algorithm _inside (A, B)_

{Let A, B be two e-complexes of dimension d $(d > 1)$, A_i, B_j the d-simplices of A and B and let $bb(A)$ and $bb(B)$ be the bounding boxes of A and B. Test if A is inside B}

Step 1:   isInside = false;
            if (not _overlap (bb(A), bb(B)))_ return isInside
            else   collect all d-simplices A_i, that have a non-empty intersection with both bounding boxes. Let QS be the set of the qualified d-simplices;

Step 2:   Search an arbitrary d-simplex A_i in QS with _inside_e_complex (A_i, B)_.
            if not found return isInside;

Step 3:   if (not _inside_e_complex (neighbours (A_i), B)_ return isInside;
            isInside = true;
            return isInside

end _inside._

In the _inside_e_complex_ function it is tested, if A_i and the neighbours of A_i, respectively, are inside B. If all three points [with tetrahedra: four points] of A_i are inside a single B_j, the triangle-inside-triangle [tetrahedron-inside-tetrahedron] problem can be reduced to a point-inside-triangle [point-inside-tetrahedron] problem. Otherwise the test must be executed recursively for the neighbours of B_j. The method is to be executed recursively for all neigh-

---

1. I.e. the bounding boxes intersect each other, but the e-complexes are disjoint.
2. For e_0- and e_1-complexes, respectively, no neighbourhoods of the simplices can be exploited. Thus it is insufficient to use traditional plane sweep techniques with not convex e_0- and e_1-complexes, respectively. For convex e_1-complexes a slightly modified algorithm can be used that exploits the neighbourhoods of the virtual 2-simplices.

bours of A_i that were not yet visited. It terminates as soon as an A_i is not inside B. This is the case, if an A_i exists for which no neighbours of B_j exist so that A_i is inside these neighbours.

The efficient realization of the *inside*-algorithm is due to the advantages of the introduced *integrated spatial representation:* These are the *"breaking down" of the algorithms to simple geometries (triangles, tetrahedra)*, the test of the *bounding boxes* and the exploitation of the *topology* of the e-complexes. The surfing on top of the topology (neighbourhood relationship) accelerated the algorithms, as it enabled the fast delimination of the relevant simplices without an a-priori sorting within the geometric operation.

Complexity of the inside-algorithm:

For *step 1* all d-simplices of A and B have to be run through twice. Thus the complexity is $O(n)$, let n be the number of d-simplices of A and B. *Step 2* has a quadratic running time concerning the number of the d-simplices from A and B, which as well intersect the bounding box of A as the bounding box of B. *Step3* has a complexity of $O(s)$, let s be the number of intersections of the d-simplices of A and B. The complexity (worst case) concerning the number n of the d-simplices which intersect both bounding boxes is $O(n^2)$.

Overlap-operators:

One problem, however, still remains: We cannot efficiently realize such operations on e-complexes, that again provide e-complexes as their result. We try to clarify this problem by means of an example. If we developed an intersection operator[1] on the basis of the above algorithms on e_2- and e_3-complexes, respectively, then this algorithm would have a crucial disadvantage: We would have to iterate over *all* triangles and tetrahedra, respectively, also if only few of the triangles and tetrahedra, respectively, intersected each other. As we can see in fig. 4.18, it can occur that *more than one connected e-complex* arises during the intersection of two e-complexes. Thus the algorithm cannot terminate as described before, if it spreads itself, starting from a start triangle A_i to triangles which do not intersect.

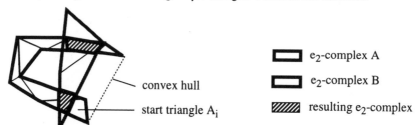

Fig. 4.18 Intersection[2] of two non-convex e_2-complexes A and B

We solve the problem by filling up the e-complexes to *convex e-complexes*. In the above example the e_2-complex A can be filled up to a convex e_2-complex. Then the algorithm can spread itself starting from a start triangle A_i towards the second intersecting surface below.

---

1. The intersection operator *overlapOp* shall provide the intersecting geometry and topology of two intersecting e-complexes.
2. The intesecting surfaces are drawn hatched.

Before we introduce the *overlap-operator* between two *convex* e-complexes (ce-complexes), we give two *cut-operators* that intersect an e_2-complex along an intersection plane. The result is an intersection line, (or more precisely: a set of connected e_1-complexes) and the triangle network which is "outcutted" inside the intersection line (or more precisely: a set of connected e_2-complexes). The difference of the algorithms for e_2- and e_3-complexes is that the intersection operations for $(d = 3)$ have to be executed on real tetrahedra instead like in the twodimensional case $(d = 2)$ on real triangles. Furthermore, the resulting geometries possibly have to be tetrahedralized instead of triangulated. Methods to be used for this we introduced with the generation of convex e-complexes (see chapter 4.4). As an auxiliary algorithm we first give the intersection of a tetrahedron with an intersection plane.

algorithm tetraCutLineOp (A, B)

{Let A be a 3-simplex (tetrahedron) and let B be an intersection plane[1]. Determine the intersection line C of A and B and return it as a convex e_1-complex (e-complex with "real" 1-simplices, i.e. edges)}

Step 1:   if (*B* is marked) <u>return</u> NULL;
              mark *B;*
              <u>if</u> <u>not</u> *overlap (A, B)* <u>return</u> *NULL;*
              new C;

Step 2:   <u>for</u> *i = 1* <u>to</u> *4*
                  C = getTriangle(i)->triCutLineOp (A, B);
                  / Intersection of the plane with the i-th triangle of A */*
                  combine (C);

Step 3:   <u>for</u> *i = 1* <u>to</u> *4*
                  C = getTetrahedron(i)->tetraCutLineOp (A, B);
                  / Intersection with the i-th neighbour tetrahedron of A */*
                  combine (C);
                  <u>return</u> *C*

end *tetraCutLineOp.*

We will later use the above algorithm as an auxiliary algorithm for intersecting e_2-complexes with a closed ring of intersecting planes.

algorithm *cutLineOp (A, B)*

{Let A be a convex e-complex with dimension $(d = 2)$ in 3D space and let $\cdot B$ be a closed ring of intersection planes (see fig. 4.19). Let A_i be a real d-simplex (triangle) of the tetrahedron T_i of A. Determine the intersection line C, which intersects A and B and return it as a convex e_1-complex (e-complex with a set of "real" simplices, i.e. edge sequences)}

Step 1:   *Bounding-box test (see above);*

Step 2:   Find a start triangle A_i of a tetrahedron T_i of A, which intersects B;

---

1. Let the intersection planes be represented by triangle networks.

Step 3:   *new C;*
                 <u>for</u> *k = 1* <u>to</u> *4*
                     <u>if</u> *overlap* (T_i, B) <u>and</u> (<u>not</u> all T_i are visited)
                         C = getTriangle(k)->triCutLineOp $(T_i, B);$
                     / Intersection with the k-th triangle of* T_i **/*

Step 4:   <u>for</u> all intersection planes <u>in</u> *B*
                     <u>for</u> *k = 1* <u>to</u> *4*
                         <u>if</u> *overlap* (T_i, B) <u>and</u> (<u>not</u> all T_i are visited)
                             $C = C \cup$ *getTetrahedron(k)->tetraCutLineOp (neighbour* $(T_i), B);$
                         / recursively for the neighbours of* T_i*, which intersect B */*
                     <u>return</u> C

end *cutLineOp.*

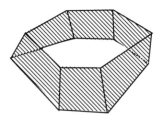

Fig. 4.19 Closed ring of intersection planes

With the method described in *tetraCutLineOp* we "surf" from the first found tetrahedron through the convex e_2-complex. Only such tetrahedra are taken into account that intersect the intersection line or that are direct neighbouring tetrahedra. If the first tetrahedron has received the intersection results from all neighbours, it combines them and returns the complete result. Fig. 4.20 shows the intersection line (real edges of the ce_1-complex *D*), that arises during the intersection of a closed ring with the ce_2-complex *A*.

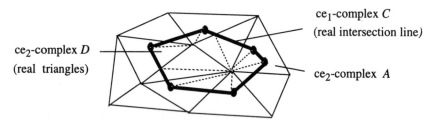

Fig. 4.20 Intersection line which arises during the intersection with a close ring of intersection planes

Complexity:

The *worst case* occurs, if no intersection plane intersects the convex e_2-complex. Let *T* be the number of tetrahedra of the convex e_2-complex and N_i the number of edges in the result of the i-th of totally *s* intersection planes. The searching of a start tetrahedron *(step 2)* needs $O(T)$ time. During the intersections with the directly involved tetrahedra and their neighbours we access on $O(N_i)$ tetrahedra *(step2 and 3)*. The combination of the line sequences needs between $O(N_i)$ and $O(N_i^2)$ time and the combination of the edges in the intersection object needs between s and $O((\sum_{i=1 \, to \, s} N_i)^2)$ time. Thus the total need for time can be estimated by an upper border of $O((\sum_{i=1 \, to \, s} (T + N_i^2)) + (\sum_{i=1 \, to \, s} N_i)^2 + T) \leq O((\sum_{i=1 \, to \, s} N_i)^2 + T)$. If each intersection plane generates a connected result and if the results are connected, the first summand approaches $(\sum_{i=1 \, to \, s} N_i) + s$ as an upper border. The summand $O(T)$ can be reduced by the use of a spatial access method like the R*-tree (BECKMANN et al. 1990).

algorithm *cutSurfaceOp (A, B)*

{Let A be a convex e-complex with dimension $(d = 2)$ in 3D space and let B be a closed ring of intersection planes. Determine the convex ce_2-complex D, whose real triangles are inside the intersection line C}

Step 1: $C = CutLineOp (A, B);$

Step 2: $D = getSurface (C);$
            /* Set of the triangles that are inside C */
            <u>return</u> D

end *cutSurfaceOp.*

In the above algorithm the intersection line with the intersection plane is determined with the *cutLineOp (Step 1)*. Then the ce-complex is intersected along the intersection line, i.e. the chosen surface is cutted out by means of the *getSurface* function from the e_2-complex *(step 2)*.

In the preparing phase for the *overlap*-algorithm after a bounding-box test, the searching for a pair of start triangles is executed. Then with the corresponding tetrahedra the actual algorithm can be started.

algorithm *prepareOverlapOp (A, B)*

{Let A, B be two convex e-complexes of dimension $(1 \leq d \leq 3)$ in 3D space and let A_i, B_j be the d-simplices (edges, triangles and tetrahedra, respectively) of A and B. Determine a pair of start d-simplices A_i, B_j of the tetrahedra T_i, T_j from A and B, that intersect each other. Return T_i and T_j}

Step 1:   Bounding-box test (see above); /* *global filter* */

Step 2:   Search a pair of d-simplices (edges, triangles and tetrahedra, respectively) A_i, B_j of the tetrahedra T_i, T_j so that *overlap* (T_i, T_j) is TRUE;   /* *local filter* */
            <u>return</u> (T_i, T_j)

end *prepareOverlapOp.*

In the *overlap* function we test, if the real d-simplices of the tetrahedra T_i and T_j intersect each other.

algorithm *overlapOp (A, B, T$_i$, T$_j$)*

{Let *A, B* be two convex e-complexes of the dimension *(1 ≤ d ≤ 3)* in 3D space and let *T$_i$, T$_j$* be the d-simplices of *A* and *B*, that were determined as the start simplices (edges, triangles and tetrahedra, respectively) in the function *prepareOverlapOp*. Determine the convex e-complex that is the intersection of *A* and *B*}

Step 1:   *new C;*
                C = computeOverlap (T$_i$, T$_j$);

Step 2:   <u>for</u> *j = 1* <u>to</u> *(d + 1)*
                <u>if</u> *overlap (T$_i$, neighbour (T$_j$))* <u>and</u> (<u>not</u> all T$_j$ are visited)
                    C = C ∪ computeOverlap (T$_i$, neighbour (T$_j$))
                /* recursive for those neighbours of *T$_j$*, that intersect *T$_i$* */
                /* "circle method", see fig. 4.21 */

Step 3:   <u>for</u> *j = 1* <u>to</u> *(d + 1)*
                <u>if</u> *overlap (neighbour (T$_i$), T$_j$)* <u>and</u> (<u>not</u> all T$_i$ are visited)
                    C = C ∪ overlapOp (A, B, neighbour (T$_i$), T$_j$);
                /* recursive for those neighbours of *T$_i$*, that intersect *T$_j$* */
                <u>return</u> *C*

end *overlapOp.*

The function *computeOverlap* computes the intersecting geometry of *T$_i$* and *T$_j$* (and their real d-simplices, respectively). The result is tetrahedralized, and, if necessary, the new neighbours of the intersecting geometries are registered.

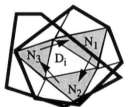

Fig. 4.21 Circle method for the searching through of a tetrahedron
           (illustrated in 2D space by a triangle T$_i$ with the neighbours N$_1$, N$_2$, N$_3$)

A big advantage of the use of convex e-complexes is that only a single pair of start simplices has to be searched. After that the algorithm spreads over the only existing intersection area of both e-complexes. It terminates automatically, if the boundary of the intersection area is reached. Note that the result of the *overlapOp*-algorithm can be of arbitrary dimension and that even resulting objects with mixed dimensions are allowed.

For e$_1$-complexes in 2D space the same algorithm as for e$_2$-complexes in 2D space is to be used after the filling up to convex e$_1$-complexes. The only difference is that the intersection is to be executed with the "real" edges instead of the triangle of an e$_2$-complex, i.e. with those edges of the e$_1$-complex that already existed before the generation of the convex hull.

For volumes whose surface is given (2 1/2-D), the same algorithm can be used, if we fill up the volume with tetrahedra (see also "$e_{2-1/2}$-complexes", chapter 6). If holes exist on the surface of the volumes, then they first have to be filled up by a triangulation. After that we can "surf along" the topology, i.e. the interior tetrahedra. Thus we can use the 2D intersection algorithm for the intersection of the surfaces. Intersection operations are only executed for the "real" triangles on the surface. Against this the tetrahedra in the interior of the volume are "virtual" tetrahedra and serve for the navigation on the topology of the e-complexes.

A problem still is the differentiation of the "interior" boundaries from the "exterior" boundary in the volumes. However, we can again mark a start triangle of which we know that it is lying on the boundary, and then run through the whole surface. This is the same procedure as we have introduced for the differentiation of interior and exterior boundaries of surfaces (see fig. 4.12). If we again meet the start triangle, then the boundary is the exterior boundary, otherwise it is an interior boundary.

Complexity:

In the worst case[1] the *overlapOp* needs $O(n^2)$ time for the *search of the start triangle (step 1 and 2 of prepareOverlapOp)*, let n be the number of triangles in the intersection box of both bounding boxes. This time, however, can be significantly reduced on average by the use of a spatial access method (see chapter 5.3).

The real intersection algorithm (*circle method, step 2 and 3 of overlapOp*) has a *linear running time relating to the number of the triangle- and the tetrahedron intersections,* respectively. This is because only such triangles and tetrahedra, respectively, are searched that either are directly involved in an intersection or that are direct neighbours of such triangles and tetrahedra, respectively. This means that a triangle and tetrahedron, respectively, need not to be further considered, if it is lying outside the circle that describes the boundary of the intersection surface and volume of both e-complexes, respectively. Thus the algorithm has the important property to be *output sensitive.*

Further algorithms

We shortly describe the functioning of the algorithms for *meet* (two e-complexes A and B touch each other from outside), *covered_by* (A touches B from inside), *equal* (boundary and interior of A and B are identical) and *disjoint.*

The *covered-* and the *meet-*algorithm terminate, if a single pair of triangles and tetrahedra, respectively, of both e-complexes fulfil the respective relationship (touching from outside and inside). After the bounding-box test, the *meet (A, B)* and *coveredBy(A, B)* algorithms, only consist of the search of a start triangle and tetrahedron, respectively (see *step 2 of the overlapOp-*algorithm), which fulfils *(meet T_i, B)* and *(coveredBy T_i, B)*, respectively. If there is such a tetrahedron T_i, then the algorithm terminates and the result is true, otherwise it is false.

---

1. Both e-complexes do not intersect, i.e. they are outside each other.

In the *equal*-algorithm, both e_2- and e_3-complexes, respectively, can be "synchronously" compared with the *circle method,* tetrahedron for tetrahedron. Two e-complexes are equal, iff all of their d-simplices are equal. Thus the running time is $O(min(m, n))$, let m and n be the number of triangles of both e-complexes. Two e-complexes are *disjoint,* if either their bounding boxes are disjoint or if the bounding boxes intersect, and no simplices of both e-complexes exist that intersect each other. In the second case, a test for intersection has to be executed for all simplices that intersect the both bounding boxes. Thus the running time is $O(n^2)$, let n be the number of the simplices that intersect both bounding boxes.

Topological 3D-algorithms

The *circle method* that was used in the *overlapOp-algorithm* is a good basis for topological algorithms in 3D space. For example it can be used to answer queries like "provide the left/ right side of an object from a given intersection line and intersection plane[1], respectively" (see fig. 4.22).

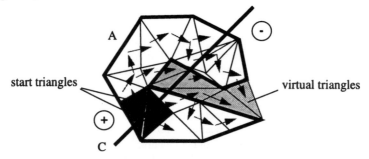

Fig. 4.22 Example of a topological 3D-operation: Positive/negative side of an e_2-complex A in
          3D space relating to an intersection plane C

algorithm *PosNegSide (A, C)*

{Input is a convex e-complex A of dimension 2 and an intersection plane *(CutSurface)* C, that cuts A into two halfs. Let P be a list of points. Determine the topology of A at one of the both sides ("positive" and "negative" side, respectively) of the intersection surface and call it T}

Step 1: $P = cutLineOp (A, C)$;

Step 2: $T = getSurface (P)$;
               <u>return</u> T

end *PosNegSide.*

Note that triangles that are lying *geometrically right* of the intersection plane can be lying *topologically left* of the intersection plane and vice versa. Thus we better call this operation *"positive/negative sideOf"* than "left/rightOf" (see fig. 4.22). In this first step the algorithm

---

1. Curved surfaces could also be used as intersection surfaces. As a limitation we only demand that the intersection surface completely intersects the convex e-complex.

determines the intersection plane on the e_2-complex and in the second step it collects the triangles on one side of the intersection plane. The definition of topological relationships often depends on the context of the application. Examples of a 3D-application from the mining can be found in (BREUNIG 1991; BREUNIG et al. 1991).

Regularized operators

Regularized operators (REQUICHA 1977; TILOVE 1980) are used to avoid "degenerated" geometries with the intersection of two areas in a plane or of two volumes in a 3D space. In the first case we intend to get an area and in the second a volume. Thus the "regularized intersection operator" *(OverlapOp*)* is defined in that way that touchings are not part of the intersection result. Thus even with geometries with holes non degenerated geometries can occur as, for instance, areas with "connected lines". The definition of the *OverlapOp*-operator* runs as follows[1] (see TILOVE 1980):

$$OverlapOp* (A, B) \quad := hull \, (^o \, (OverlapOp \, (A, B)))$$
$$= (^o \, (OverlapOp \, (A, B))) \cup @ \, (^o \, (OverlapOp \, (A, B)))$$

Thus the *regularization of a geometry* is defined as the hull of its interior points. The hull is the union of the interior and the boundary. As the simplices are again defined on the "boundary" and the "interior", the regularization is implicitly presupposed in the e-complex. The intersection of two e_2-complexes in the plane, for instance, only returns the value *true,* if their boundaries *and* their interiors intersect.

The problem of imprecise geometries

The problem of imprecise geometries, caused by finite resolution of the representation of real-numbers in the computer, cannot be reinforced in this context. We refer to the approach of the so called "Realms" by GÜTING & SCHNEIDER (1993). In this approach the coordinate of the geometries are moved on a grid before and after the application of geometric algorithms to map them on integer values. A corresponding solution is likely to be realized on a lower level of our model by mapping the coordinates of the 0-simplices (points) on a grid.

4.7 Foundations of the ECOM-Algebra

In the prior chapter efficient geometric algorithms for e-complexes were introduced. We now intend to have a closer look to the *specification of spatial operations on e-complexes.* For that the *ECOM-Algebra,* an algebra with convex e-complexes (ECom), connected sub-e-complexes (SECom) and boxes[2] as sorts S_{ECOM} and spatial operations in 3D space are introduced.

$$S_{ECOM} = \{ECom, SECom, box\}$$

---

1. @ means "boundary" and ^{o} stands for "interior".
2. Cubes that are directed to the axes of the coordinate system.

On the ECOM-algebra the following types of spatial operations are defined:

<u>Topological and
direction operations:</u>

- topological relationships	ECom	×	ECom	→ bool	(global)
and direction relationships:	SECom	×	SECom	→ bool	(local)
- topological operators:	ECom	×	ECom	→ ECom	(global)
	SECom	×	SECom	→ SECom	(local)

<u>metrical operators:</u>

| | ECom | × | ECom | → real | (global) |
| | SECom | × | SECom | → real | (local) |

<u>geometric properties:</u>

| | ECom | | | → real | (local) |

<u>geometric cut- and paste-
operators:</u>

| | ECom | × | box | → SECom | (local) |
| | SECom × ECom × box | | | → ECom | (local) |

Let SECom be a subset of connected simplices of an e-complex. We decide that sub-complexes of a single e-complex may overlap themselves. The intersection always consists of a subset of the simplices of the e-complex and can be directly deduced from the topology. Then the intersection exactly consists of the simplices which occur in the corresponding sub-complexes. Thus an operation between sub-complexes ("local" operation of an e-complex) is defined on the simplices of a single e-complex.

An e-complex (ECom) may consist of a set of non-connecting sub-geometries and topologies, respectively. We intend to guarantee that the result of set operations (for example, intersection) on e-complexes is again an e-complex (closed set of objects). Thus the ECOM-algebra additionally provide the following internal, auxiliary operators that are only defined for e_2- and e_3-complexes:

triang2D:	polygon2D	→ E_2Com, in 2D space
triang3D:	polygon3D	→ E_2Com, in 2D space
tetra3D:	polyhedron3D	→ E_3Com, in 3D space

triang2D transfers a polygon in 2D space into an e-complex of dimension 2 *(E_2Com)*, i.e. it triangulates the polygon and registers the neighbourhoods of the triangles. Correspondingly, *triang3D* transfers a polygon in 3D space into an e_2-*complex*. *tetra3D* transfers a polyhedron into an e-complex of dimension 3 *(E_3Com)*, i.e. it tetrahedralizes the polyhedron and registers the neighbourhoods of the tetrahedra.

Let us introduce the different spatial operations of the ECOM-algebra. As introduced in chapter 3.3.3, we specify topological relationships by the topological concepts *"boundary"* and *"interior"*. Finally we go into local operations, the *geometric properties* and the *geometric cut- and paste-operators*[1].

---

1. To use the operations of the ECOM-algebra in a 3D-GIS, they could be embedded into a descriptive, object-oriented query language like O_2SQL (BANCILLHON et al. 1989).

4.7.1 Topological Relationships

We distinguish 10 main groups of topological relationships on e-complexes with regard to the dimension of the participating geo-objects.

dimension 1st geo-object → ————————— dimension 2nd geo-object ↓	0	1	2	3
0	1)	-	-	-
1	2)	5)	-	-
2	3)	6)	8)	-
3	4)	7)	9)	10)

Table 4.2 The 10 main groups of topological relationships on e-complexes

We first give a basis set of topological relationships between *connected e-complexes* for each of the 10 main groups (see table 4.2), respectively[1].

Definition 4.13: An e-complex E of dimension 0 is called *connected,* if it consists of exactly one 0-simplex.

We do not only consider different topological relationships between e-complexes, but we also distinguish between different *configurations of topological relationships* within one relationship on "simplex level". This means that we distinguish the topological configurations by their simplices that touch or intersect themselves, respectively. Thus the intersection result of the boundary and the interior of the local, i.e. the *internal topology of the e-complexes* is analyzed (see also table 4.3 - 4.9). This enables an interesting differentiation of topological relationships that can be used for the identification of fine differences of the position. Thus the user may choose special configurations of the relationships, which go beyond the general relationships like *equal, disjoint, meet and inside* etc. The differentiated consideration of topological relationships is also of interest, if the topology of the e-complexes is connected with thematic information. For instance, with regard to a certain thematic attribute for an interpolation on a geological surface it can be of importance if a considered point lies inside or on the boundary of a triangle of another e-complex. If two e-complexes are touching in an edge (1-simplex), then this relationship can be specified with the intersection of the boundaries and interiors of two 1-simplices (see table 4.4). An example from the geosciences is the definition of so called fault scarps, i.e. two surfaces in 3D space that touch each other in an edge.

---

1. The main groups for 1D and 2D space follow HOOP et al. (1992), who, however, only have examined topological relationships for single line segments and areas. Connected e-complexes of dimension (d > 0) we have already introduced in definition 4.7.

Following EGENHOFER (1989) we rename the relationship so far designated as *exclusion* and *outside*-relationship, respectively, and call it *disjoint*-relationship from now on.

1) e_0-complex/e_0-complex:

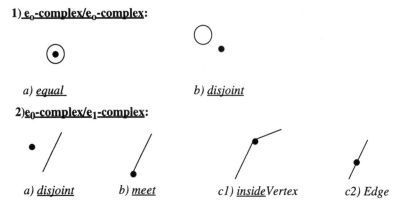

a) *equal*   b) *disjoint*

2) e_0-complex/e_1-complex:

a) *disjoint*   b) *meet*   c1) *insideVertex*   c2) *Edge*

Fig. 4.23 Topological relationships for the main groups 1 and 2

In the first main group only two topological relationships occur, because two connected e_0-complexes (points) only can be equal or not equal[1]. In the second main group already four different relationships occur. A connected e_0-complex and e_1-complex are *disjoint,* if the 0-simplex of the e_0-complex is outside the e_1-complex. Note that this is the first time that the internal topological structure of the e-complex is used, i.e. we can distinguish *insideVertex* from *insideEdge*.

3) e_0-complex/e_2-complex:

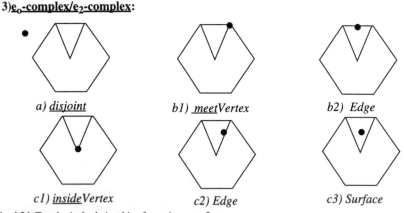

a) *disjoint*   b1) *meetVertex*   b2) *Edge*

c1) *insideVertex*   c2) *Edge*   c3) *Surface*

Fig. 4.24 Topological relationships for main group 3

In the third main group we distinguish *disjoint, meet* and *inside*. If the e_0-complex lies on a vertex, an edge and a triangle of the e_1-complex, respectively, we distinguish between relationships of a *Vertex,* an *Edge* or a triangle *(Surface).*

---

1. In operation 1a) the small black filled circles (0-simplices) belong to the first e_0-complex and the larger, not filled circles (as well 0-simplices) belong to the second e_0-complex. The equality is indicated by concentric circles.

4) e_0-complex/e_3-complex:

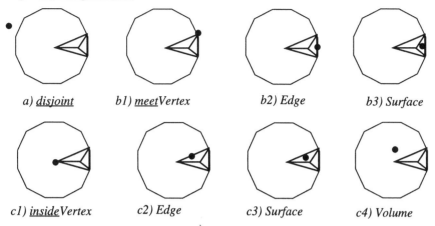

     a) disjoint   *b1) meetVertex*   *b2) Edge*   *b3) Surface*

     c1) insideVertex   *c2) Edge*   *c3) Surface*   *c4) Volume*

Compared with the third main group, in the fourth main group the relationships *meet-Surface* and *insideVolume* are added.

Fig. 4.25 Topological relationships for main group 4

5)e_1-complex/e_1-complex:

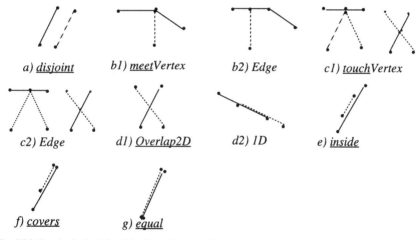

     a) disjoint   *b1) meetVertex*   *b2) Edge*   *c1) touchVertex*

     c2) Edge   *d1) Overlap2D*   *d2) 1D*   *e) inside*

     f) covers   *g) equal*

Fig. 4.26 Topological relationships for main group 5

In the fifth main group, the difference between *meet* and *touch* has to be taken into consideration. Whereas the start vertex (end vertex, respectively) of the first line sequence meets the second line sequence in its interior in the *meet*-relationship, in the *touch*-relationship both line sequences touch only with their interiors[1]. In the *overlap*-relationship we distinguish the overlapping in *one* (1D) and in *two* dimensions (2D). In 3D space it is not useful

---

1. For a line sequence its start and end vertex are its boundary, whereas the remaining parts of the geometry in between are the interior of the line sequence.

to distinguish, if the first line sequence is "crossing" the second or not. I.e. there is no reason to distinguish an angle of 180 degrees between the line sequences from the other cases (see fig. 4.26 *c1* and *c2*). In 2D space, however, it could be useful to distinguish, if the first line sequence crosses the second one or if it only touches it. The *covers*-relationship describes the touching of the boundaries, i.e. of the start and the end vertex of the line sequence, respectively. For *meet* and *touch* we internally distinguish if the e_1-complexes touch at a *Vertex* or an *Edge*.

6) e_1-complex/e_2-complex:

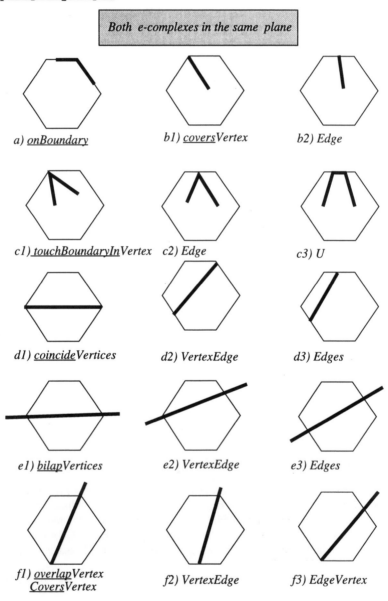

a) onBoundary	*b1) coversVertex*	*b2) Edge*
c1) touchBoundaryInVertex	*c2) Edge*	*c3) U*
d1) coincideVertices	*d2) VertexEdge*	*d3) Edges*
e1) bilapVertices	*e2) VertexEdge*	*e3) Edges*
f1) overlapVertex *CoversVertex*	*f2) VertexEdge*	*f3) EdgeVertex*

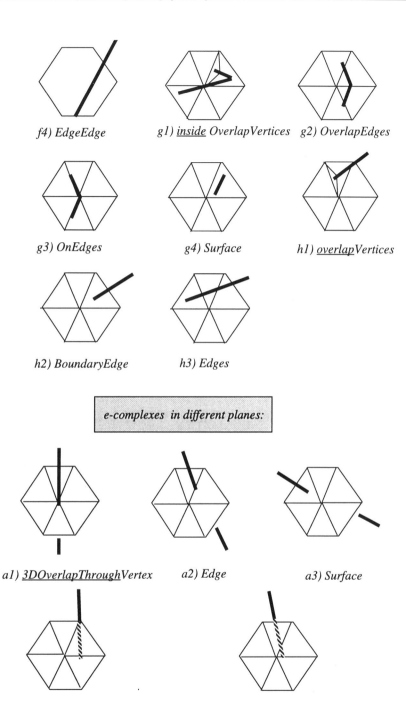

f4) EdgeEdge   *g1) inside OverlapVertices*   *g2) OverlapEdges*

g3) OnEdges   *g4) Surface*   *h1) overlapVertices*

h2) BoundaryEdge   *h3) Edges*

e-complexes in different planes:

a1) 3DOverlapThroughVertex   *a2) Edge*   *a3) Surface*

b1) 3DOverlapThroughBoundaryVertex   *b2) Edge*

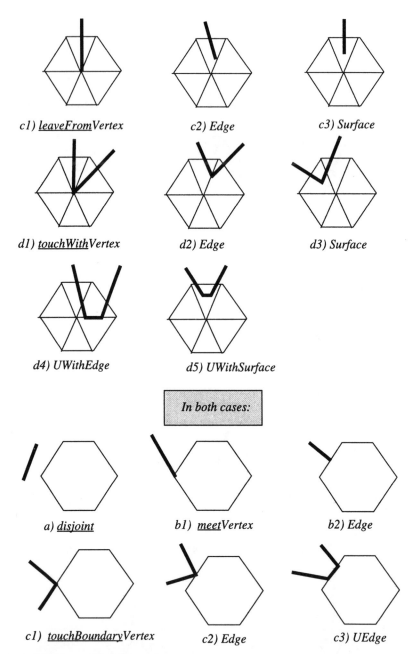

cl) *leaveFromVertex*   c2) Edge   c3) Surface

dl) *touchWithVertex*   d2) Edge   d3) Surface

d4) UWithEdge   d5) UWithSurface

In both cases:

a) *disjoint*   bl) *meetVertex*   b2) Edge

cl) *touchBoundaryVertex*   c2) Edge   c3) UEdge

Fig. 4.27 Topological relationships for main group 6

In the sixth main group we distinguish, if the e-complexes are lying in the same plane or not. For the third category of relationships it is indifferent, if the e-complexes are lying in one plane or not.

7)e₁-complex/e₃-complex:

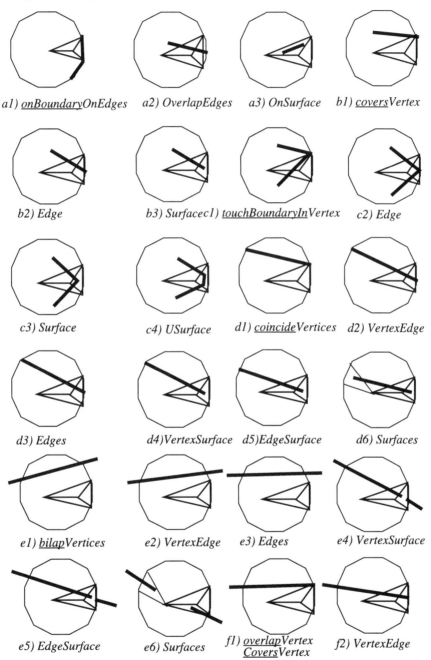

a1) onBoundaryOnEdges   _a2) OverlapEdges_   _a3) OnSurface_   _b1) coversVertex_

b2) Edge   _b3) Surfacec1) touchBoundaryInVertex_   _c2) Edge_

c3) Surface   _c4) USurface_   _d1) coincideVertices_   _d2) VertexEdge_

d3) Edges   _d4)VertexSurface_   _d5)EdgeSurface_   _d6) Surfaces_

e1) bilapVertices   _e2) VertexEdge_   _e3) Edges_   _e4) VertexSurface_

e5) EdgeSurface   _e6) Surfaces_   _f1) overlapVertex CoversVertex_   _f2) VertexEdge_

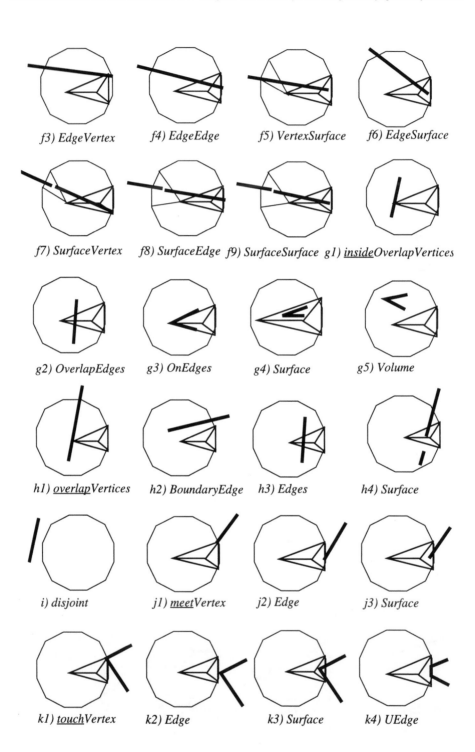

f3) EdgeVertex   f4) EdgeEdge   f5) VertexSurface   f6) EdgeSurface

f7) SurfaceVertex   f8) SurfaceEdge   f9) SurfaceSurface   g1) <u>inside</u>OverlapVertices

g2) OverlapEdges   g3) OnEdges   g4) Surface   g5) Volume

h1) <u>overlap</u>Vertices   h2) BoundaryEdge   h3) Edges   h4) Surface

i) disjoint   j1) <u>meet</u>Vertex   j2) Edge   j3) Surface

k1) <u>touch</u>Vertex   k2) Edge   k3) Surface   k4) UEdge

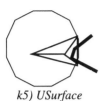

k5) USurface

Fig. 4.28 Topological relationships for main group 7

In the seventh main group we do not have to distinguish the position in a plane. Because of the three dimensions of the objects only such relationships occur that refer to the surface or the volume of the objects.

8) e₂-complex/e₂-complex:

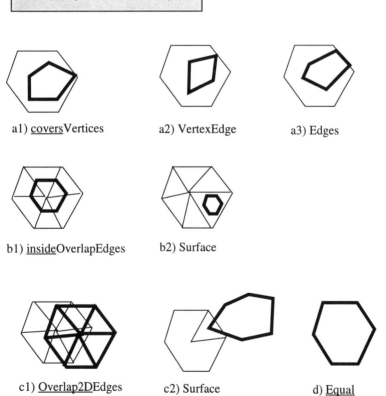

| Both e-complexes in the same plane |

a1) coversVertices   a2) VertexEdge   a3) Edges

b1) insideOverlapEdges   b2) Surface

c1) Overlap2DEdges   c2) Surface   d) Equal

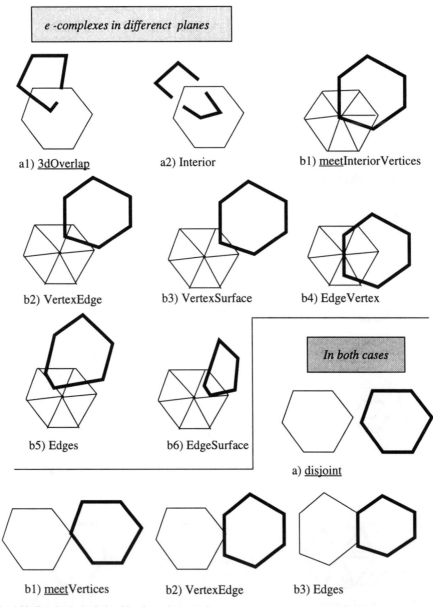

Fig. 4.29 Topological relationships for main group 8

In the eigth main group we again have to distinguish between relationships in the same plane and in different planes. Like in the sixth main group also relationships exist that are valid for both cases.

9)e$_2$-complex/e$_3$-complex:

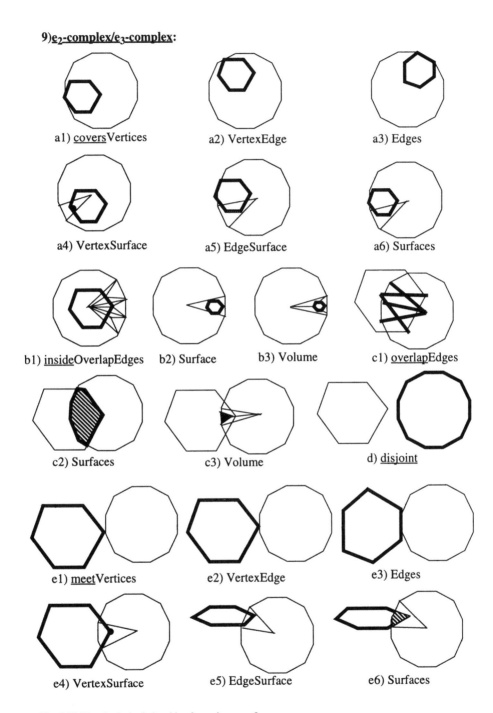

a1) <u>covers</u>Vertices   a2) VertexEdge   a3) Edges

a4) VertexSurface   a5) EdgeSurface   a6) Surfaces

b1) <u>inside</u>OverlapEdges   b2) Surface   b3) Volume   c1) <u>overlap</u>Edges

c2) Surfaces   c3) Volume   d) <u>disjoint</u>

e1) <u>meet</u>Vertices   e2) VertexEdge   e3) Edges

e4) VertexSurface   e5) EdgeSurface   e6) Surfaces

Fig. 4.30 Topological relationships for main group 9

Main group 9 has substantially less relationships than main group 8, as no differentiation concerning the plane is necessary. In the *overlap*-relationship we distinguish, if the edges of the e_2- and the e_3-complexes intersect *(Edges)*, if the intersection line is exactly lying on surfaces of the e_3-complex *(Surfaces)* or if the e_2-complex only intersects itself with a single tetrahedron of the e_3-complex *(Volume)*. Comparable, the *insideOverlap*-relationship can also be devided into these three cases. In this relationship the e_2-complex is inside the e_3-complex and intersects the interior simplices of the e_3-complex. *Meet* and *covers* have the same meaning as in the previous main groups.

10) e₃-complex/e₃-complex:

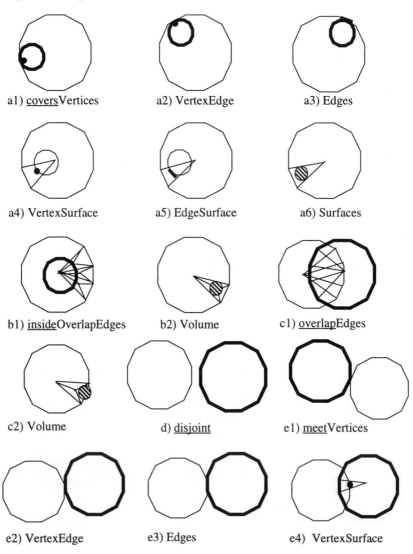

a1) coversVertices   a2) VertexEdge   a3) Edges

a4) VertexSurface   a5) EdgeSurface   a6) Surfaces

b1) insideOverlapEdges   b2) Volume   c1) overlapEdges

c2) Volume   d) disjoint   e1) meetVertices

e2) VertexEdge   e3) Edges   e4) VertexSurface

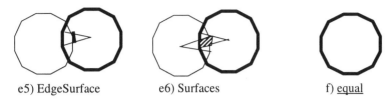

e5) EdgeSurface   e6) Surfaces   f) <u>equal</u>

Fig. 4.31 Topological relationships for main group 10

The names of the relationships for the main group 10 are nearly the same as for main group 9. In the *overlap*-relationship we distinguish, if the *Edges* of the e_3-complexes intersect each other or if one of the two e_3-complexes intersects the other one only in a single tetrahedron *(Volume)*. In main group 10, the *equal*-relationship is added as before in the main groups 1, 5 and 8, because both involved objects have the same dimension.

Many other topological relationships may be derived from the introduced relationships between connected e-complexes. Ths the relationships of the main group 1 to 10 are to be seen as *basic relationships* from which combinations of more complex relationships can be derived.

Specification of topological relationships

In chapter 3.3.3 we have introduced the specification of topological relationships by Egenhofer, based on the topological concepts of "boundary @" and "interior" for objects in 2D space and with codimension[1] 0. The topological specification avoids unnecessary implementation details like explicit coordinates. EGENHOFER (1989b) proved that eight relationships for two d-simplices *(d > 1)* with codimension 0 are existing that can be specified by the binary values "empty $\varnothing$" and "not empty $\neg \varnothing$", refering the intersections of their boundaries and their interiors. These 8 relationships are *disjoint, meet, overlap, inside, contains, covers, coveredBy* and *equal*. As Egenhofer showed, the relationships *contains* and *coveredBy* are redundant.

Let us extend the specification for objects of arbitrary codimension. As we have seen in the 10 main groups before, it does not suffice to consider the intersection of the boundaries and the interiors of the *e-complexes* to distinguish the topological *configurations* of one relationship. Thus the *simplices* have to be considered "one level below", because they are relevant for the result of the respective topological configuration. At this point it also should become clear, why the internal topology of the objects is also relevant for *global* queries (see chapter 4.2) and why it is ingenious to consider the same structure of e-complexes for both, local and global queries. In the following we demonstrate the specification of the topological relationships for e_0-, e_1-, e_2- and e_3-complexes with codimension 0. We extend them by topological configurations, which are derived by the internal topology of the e-complexes. Finally, we give an example for specifications with objects of arbitrary codimension, i.e. for the main groups 1 to 10, for the *inside*-relationship.

---

1. The codimension is the difference between the dimension of the objects and the considered space.

Main group 1 (e_0-complex/e_0-complex):

Relationship	$@\cap@$	$@\cap^\circ$	$^\circ\cap@$	$^\circ\cap^\circ$
equal	$\neg\varnothing$	$\neg\varnothing$	$\neg\varnothing$	$\neg\varnothing$
disjoint	$\varnothing$	$\varnothing$	$\varnothing$	$\varnothing$

Table 4.3 Specification of topological relationships for e_0-complexes

Main group 5 (e_1-complex/e_1-complex[1]):

Relationship	Comparison on level of	$@\cap@$	$@\cap^\circ$	$^\circ\cap@$	$^\circ\cap^\circ$
disjoint	1-complexes				
meetVertex	1-complexes		$\neg\varnothing$		
	1-simplices	$\neg\varnothing$			
Edge	1-complexes		$\neg\varnothing$		
	1-simplices		$\neg\varnothing$		
touchVertex	1-complexes				$\neg\varnothing$
	1-simplices	$\neg\varnothing$			
Edge	1-complexes				$\neg\varnothing$
	1-simplices		$\neg\varnothing$		
overlap2D	1-complexes				$\neg\varnothing$
1D	1-complexes		$\neg\varnothing$	$\neg\varnothing$	$\neg\varnothing$
overlapMeet **Vertex**	1-complexes				$\neg\varnothing$
	1-simplices		$\neg\varnothing$		
Vertices	1-complexes				$\neg\varnothing$
	1-simplices	$\neg\varnothing$			
inside	1-complexes		$\neg\varnothing$		$\neg\varnothing$
covers	1-complexes	$\neg\varnothing$	$\neg\varnothing$		$\neg\varnothing$
equal	1-complexes	$\neg\varnothing$			$\neg\varnothing$

Table 4.4 Specification of topological relationships for e_1-complexes

---

1. In this and in the following tables the first line of each field in the right side of the table is corresponding to the comparison of the objects on the "complex level", whereas the second line considers the comparison on the "simplex level". To simplify the contents of the tables, the sign "$\varnothing$" was omitted for those cases in which the result of the intersection is the empty set. Thus the empty fields of this and the following tables stand for "$\varnothing$", i.e. the empty set.

Main group 8 (e_2-complex/e_2-complex):

In one plane:

Relationship	Comparison on level of	$@\cap@$	$@\cap°$	$°\cap@$	$°\cap°$
covers **Vertices**	2-complexes 1-simplices	¬ ∅ ¬ ∅	¬ ∅		¬ ∅
VertexEdge	2-complexes 1-simplices	¬ ∅	¬ ∅ ¬ ∅		¬ ∅
Edges	2-complexes 1-simplices	¬ ∅	¬ ∅ ¬ ∅		¬ ∅ ¬ ∅
inside **OverlapEdges**	2-complexes 1-simplices	¬ ∅	¬ ∅ ¬ ∅		¬ ∅ ¬ ∅
Surface	2-complexes 1-simplices		¬ ∅ ¬ ∅		¬ ∅ ¬ ∅
Overlap2D **Edges**	2-complexes 1-simplices	¬ ∅	¬ ∅	¬ ∅	¬ ∅
Surface	2-complexes 1-simplices	¬ ∅	¬ ∅	¬ ∅	¬ ∅
equal	2-complexes 1-simplices	¬ ∅			¬ ∅

Table 4.5 Specification of topological relationships for e_2-complexes in a plane

In the configurations of the *covers*-relationship we differentiate, if the two e_2-complexes touch each other by two *Vertices,* by a vertex and an edge *(VertexEdge)* or by two *Edges.* For the *inside-* and the *overlap2D*-relationship we distinguish, if only a single or several 2-simplices (triangles) are relevant for the relationship.

Not in one plane:

Relationship	Comparison on level of	$@\cap@$	$@\cap°$	$°\cap@$	$°\cap°$
3DOverlap	2-complexes		¬ ∅	¬ ∅	¬ ∅
Interior	2-complexes		¬ ∅		¬ ∅
meetInterior Vertices	2-complexes 1-simplices	 ¬ ∅	¬ ∅		
VertexEdge	2-complexes 1-simplices		¬ ∅ ¬ ∅		
VertexSurface	2-complexes 1- and 2- simplices		¬ ∅ ¬ ∅		
EdgeVertex	2-complexes 1-simplices		¬ ∅	 ¬ ∅	
Edges	2-complexes 1-simplices		¬ ∅	 ¬ ∅	 ¬ ∅
EdgeSurface	2-complexes 1- and 2-simplices		¬ ∅ ¬ ∅		 ¬ ∅

Table 4.6 Specification of topological relationships for e_2-complexes in different planes

The both cases of the *3DOverlap*-relationship can be distinguished at the complex-level: For the first case holds that boundary and interior of both e_2-complexes and the two interiors intersect each other. In the second case, however, only the boundary of the first e_2-complex and the interior of the second e_2-complex and both interiors intersect. To distinguish this case from the *meet*-relationship, the intersection of the 1-simplices has to be examined. Similiar tests are necessary for those configurations of the *meet*-relationships for which it is no matter if the e-complexes are lying in the same plane or not (see table 4.7).

In both cases:

Relationship	Comparison on level of	$@\cap@$	$@\cap°$	$°\cap@$	$°\cap°$
disjoint	2-complexes				
meet **Vertices**	2-complexes 1-simplices	¬ ∅ ¬ ∅			
VertexEdge	2-complexes 1-simplices	¬ ∅ ¬ ∅	¬ ∅		
Edges	2-complexes 1-simplices	¬ ∅ ¬ ∅			¬ ∅

Table 4.7 Specification of topological relationships for e_2-complexes in the same and in different planes

Main group 10 (e_3-complex/e_3-complex)

Relationship	Comparison on level of	$@\cap@$	$@\cap°$	$°\cap@$	$°\cap°$
coversVertices	3-complexes 1-simplices	¬ ∅ ¬ ∅	¬ ∅		¬ ∅
VertexEdge	3-complexes 1-simplices	¬ ∅	¬ ∅ ¬ ∅		¬ ∅
Edges	3-complexes 1-simplices	¬ ∅ (¬ ∅)	¬ ∅ (¬ ∅)		¬ ∅ ¬ ∅
VertexSurface	3-complexes 1- and 2-simplices	¬ ∅	¬ ∅ ¬ ∅		¬ ∅
EdgeSurface	3-complexes 1- and 2-simplices	¬ ∅	¬ ∅ (¬ ∅)		¬ ∅ ¬ ∅
Surfaces	3-complexes 2-simplices	¬ ∅ (¬ ∅)	¬ ∅		¬ ∅ ¬ ∅
insideOverlap **Edges**	3-complexes 2-simplices	¬ ∅	¬ ∅ ¬ ∅		¬ ∅ ¬ ∅
Volume	3-complexes 2-simplices		¬ ∅		¬ ∅
overlapEdges	3-complexes 2-simplices	¬ ∅ ¬ ∅	¬ ∅ ¬ ∅	¬ ∅ ¬ ∅	¬ ∅ ¬ ∅
Volume	3-complexes 2-simplices	¬ ∅	¬ ∅	¬ ∅	¬ ∅

Relationship	Comparison on level of	@∩@	@∩°	°∩@	°∩°
disjoint	3-complexes				
	2-simplices				
meet Vertices	3-complexes	¬∅			
	1-simplices	¬∅			
VertexEdge	3-complexes	¬∅			
	1-simplices		¬∅		
Edges	3-complexes	¬∅			
	1-simplices	(¬∅)	(¬∅)		¬∅
VertexSurface	3-complexes	¬∅			
	1- and 2-simplices		¬∅		
EdgeSurface	3-complexes	¬∅			
	1- and 2-simplices		¬∅		¬∅
Surfaces	3-complexes	¬∅			
	2-simplices	(¬∅)			¬∅
equal	3-complexes	¬∅			¬∅

Table 4.8 Specification of topological relationships for e_3-complexes

For a closer examination of the two configurations of the *overlap*-relationships, it has to be tested if the first e_3-complex intersects 2-simplices of the second e_3-complex which are not part of its boundary. The same thing is valid for *insideOverlap*. In the case of touching of vertices or edges *(Vertices, VertexEdge, Edges)* in the *meet*-relationships, the intersections of 1-simplices have to be tested. Otherwise the intersection of 1- and 2-simplices, respectively, have to be tested *(VertexSurface, EdgeSurface, Surfaces)*. The same procedure is used for the *covers*-relationship. In the *meetEdges, meetSurfaces* and the *coversEdges* relationships, either both boundaries or one boundary of the first simplex with the interior of the second simplex are intersecting besides the interiors of the 1-simplices (2-simplices) . Thus in table 4.8 these exclusive cases are always set into brackets.

We have seen in the tables 4.3 - 4.8 how the different topological relationships for connected e-complexes can be specified with codimension 0. In the following, we give the specification for objects of arbitrary codimension (main groups 1 - 10) by the example of the *inside*-relationship.

Relationship	Main group	Comparison on level of	@∩@	@∩°	°∩@	°∩°
equal	1	0-complexes	¬ ∅			¬ ∅
inside	2	0-,1-complexes		¬ ∅		¬ ∅
Vertex		0-,1-simplices	¬ ∅		¬ ∅	
Edge		0-,1-simplices		¬ ∅		¬ ∅
inside	3	0-,2-complexes		¬ ∅		¬ ∅
Vertex		0-,1-simplices	¬ ∅		¬ ∅	
Edge		0-,1-simplices		¬ ∅		¬ ∅
Surface		0-,2-simplices		¬ ∅		¬ ∅
inside	4	0-,3-complexes		¬ ∅		¬ ∅
Vertex		0-,1-simplices	¬ ∅		¬ ∅	
Edge		0-,1-simplices		¬ ∅		¬ ∅
Surface		0-,2-simplices		¬ ∅		¬ ∅
Volume		0-,3-simplices		¬ ∅		¬ ∅
inside	5	1-complexes		¬ ∅		¬ ∅
inside	6	1-,2-complexes		¬ ∅		¬ ∅
Overlap Vertices		0-,1-simplices			¬ ∅	¬ ∅
Overlap Edges		1-simplices				¬ ∅
OnEdges		1-simplices		¬ ∅	¬ ∅	¬ ∅
Surface		1-,2-simplices		¬ ∅		¬ ∅
inside	7	1-,3-complexes		¬ ∅		¬ ∅
Overlap Vertices		1-simplices			¬ ∅	
Overlap Edges		1-simplices				¬ ∅
OnEdges		1-simplices	¬ ∅			¬ ∅
Surface		1-,2-simplices		¬ ∅		¬ ∅
Volume		1-,3-simplices		¬ ∅		¬ ∅

Relation-ship	Main group	Comparison on level of	@∩@	@∩°	°∩@	°∩°
inside	8	2-complexes		¬ ∅		¬ ∅
Overlap Edges		1-simplices	¬ ∅	¬ ∅		¬ ∅
Surface		1-simplices		¬ ∅		¬ ∅
inside	9	2-,3-complexes		¬ ∅		¬ ∅
Overlap Edges		1-simplices				¬ ∅
Surface		2-simplices		¬ ∅		¬ ∅
Volume		2-,3-simplices		¬ ∅		¬ ∅
inside	10	3-complexes		¬ ∅		¬ ∅
Overlap Edges		1-simplices				¬ ∅
Volume		3-simplices		¬ ∅		¬ ∅

Table 4.9 Specification of the *inside*-relationship for connected e-complexes of arbitrary codimension

Table 4.9 shows the specification of the *inside*-relationship for e-complexes of arbitrary co-dimension. As there is no "inside"-relationship for main group 1, we use the "equal"-relationship for main group 1 instead. As we have seen, the specification on the intersection of the boundaries and interiors of simplices can also be applied for relationships between objects of arbitrary dimension.

Extension of the topological relationships for general, not connected e-complexes

The above specifications are valid for binary relationships between *connected and convex e-complexes*. To extend them to general, not connected e-complexes, we define set-valued relationships.

Let $A := \{A_1, ..., A_m\}$ and $B := \{B_1, ..., B_n\}$ be two sets of connected and convex e-complexes with $i, j \in 1, ..., m, k \in 1, ..., n$.

Definition 4.14 *(set-valued overlap ∩):* A set of connected and convex e-complexes *A over-laps* a set of connected and convex e-complexes *B* , if an e-complex $A_i \in A$ exists that over-laps at least one e-complex $B_k \in B$:

$$A \cap B \Leftrightarrow \exists\, A_i \in A : A_i \cap B_k.$$

Definition 4.15 *(set-valued meet* ⊥*):* A set of connected and convex e-complexes *A touches* a set of connected and convex e-complexes *B* from outside *(meets),* iff an e-complex $A_i \in A$ exists that touches an e-complex $B_j \in B$ from outside and if it does not intersect other e-complexes from *B* and if additionally all other e-complexes from *A* touch at least one e-complex from *B* from outside or if they are disjoint to each $B_k \in B$:

$$A \perp B \Leftrightarrow \exists A_i \in A, B_k \in B : A_i \perp B_k \land \neg (A_i \cap B_k) \land$$
$$\forall A_j \backslash A_i, B_l \backslash B_k : (A_j \perp B_l) \lor (A_j \parallel B_l).$$

Definition 4.16 *(set-valued covers* ⊇*):* A set of connected and convex e-complexes *A covers* a set of connected and convex e-complexes *B*, iff an e-complex $A_i \in A$ exists that covers all B_k or iff a subset of *A* exists so that it covers all $B_k \in B$:

$$A \supseteq B \Leftrightarrow \exists A_i \in A : A_i \supseteq B \lor \exists_{i1, ..., ij} : \cup(A_{i1}, ..., A_{ij}) \supseteq B.$$

Definition 4.17 *(set-valued inside* ⊂*):* A set of connected and convex e-complexes *A* is *inside* a set of connected and convex e-complexes *B*, iff a $B_k \in B$ exists so that all $A_i \in A$ are covered by the $B_k \in B$:

$$A \subset B \Leftrightarrow \exists B_k \in B : \forall A_i \in A : A_i \subset B_k.$$

Definition 4.18 *(set-valued disjoint* ∥*):* A set of connected and convex e-complexes *A* is *disjoint* to a set of connected and convex e-complexes *B*, iff all e-complexes $A_i \in A$ are outside all $B_k \in B$:

$$A \parallel B \Leftrightarrow \forall A_i \in A, B_k \in B : A_i \parallel B_k.$$

Definition 4.19 *(set-valued equal* =*):* A set of connected and convex e-complexes *A* is *equal* to a set of connected and convex e-complexes *B*, iff for each $A_i \in A$ a $B_k \in B$ exists that is equal to A_i:

$$A = B \Leftrightarrow \forall A_i \in A : \exists B_k : A_i = B_k.$$

4.7.2 Direction Relationships

In the following we will include such direction relationships into the ECOM-algebra that can be determined straight forwardly. First we restrict ourselves to convex e-complexes and secondly we demand that the e-complexes must not be "too near together". This means that the geometries of the objects have to be spatially separated in a way that an unambigious determination of direction relationships is provided. It does not seem to be useful to carry out a specification of direction relationships for not-convex e-complexes, because we would have to consider too many special cases. In fig. 4.31, for instance, we cannot distinguish if *A* is *northOf B* or if *B* is *northOf A*. Thus we restrict ourselves to direction relationships between connected and convex e-complexes (ce-*complexes*)[1].

---

1. Direction relationships between concave and not connected e-complexes can be determined by the help of the generation of the convex hull of the e-complexes.

Fig. 4.32 Unambigious determination of the direction relationship for not convex e-complexes

For the specification of the direction relationships in 2D space, we use a simple raster approach, i.e. we take into consideration eight raster cells around the bounding box of an e-complex A. Those give the direction in which a second e-complex B is lying relating to A (see fig. 4.33).

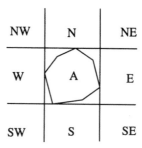

Fig. 4.33 Direction relationships on e-complexes in 2D space

For the moment we choose a "conservative" method to determine the directions. In fig. 4.34 a) A is lying *northOf B*, because *every point* of A is lying *northOf B*. We call this approach the "point rule".

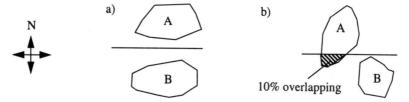

Fig. 4.34 a) A is lying *northOf B* according to the "point rule"
         b) A is lying *northOf B* according to the "percentage rule"

The "point rule", however, can be weakened, if we also allow small overlappings beyond the raster borders. We introduce a percentage tag (10% in our example) for an e-complex, up to which an overlapping is allowed respective to the coordinate area $(B_{xmin}, B_{xmax} \mid B_{ymin}, B_{ymax})$ of its corresponding raster cell. We call this approach the "percentage rule". In fig. 4.34 b) A lies *northOf B*, according to the percentage rule, as for only 10% of the points the point rule is not valid, i.e. only 10% of the area of A are projected into the raster cell of B.

The point rule and the percentage rule can be transfered straight forwardly to the threedimensional space. Consider a cube-like section of the threedimensional Euclidean space, which is subdivided into 27 cells. Then, altogether 26 different _basic direction relationships_ can be specified between a convex e-complex (ce-complex) B and a convex e-complex A whose spatial location is in the centre of the given space (see fig. 4.35).

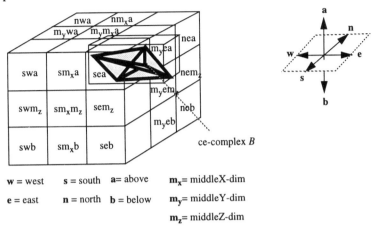

w = west	s = south	a= above	m_x= middleX-dim
e = east	n = north	b = below	m_y= middleY-dim
			m_z= middleZ-dim

Fig. 4.35 Direction relationships on ce-complexes in 3D space

By the combinations of the basic direction relationships, however, more detailed direction relationships can be determined. In fig. 4.35, for example, the following relationship between the ce-complexes[1] A and B is valid:

$$(B\ sea\ A) \wedge (B\ m_yea\ A) \implies (B\ e\ A) \wedge \neg(B\ n\ A) \wedge (B\ a\ A),$$

i.e. B is _eastOf A_ and _above A_, or shorter: B is _eastOf_ and _southeastOf A_ and _above A_. The 26 possible direction relationships can be derived from the combinations of the six direction relationships[2] of table 4.10 being specified by absolute coordinates.

Direction relationship	Geometric specification
A westOf B	$Xmax_{bb(A)} < Xmin_{bb(B)}$
A northOf B	$Ymin_{bb(A)} < Ymax_{bb(B)}$
A eastOf B	$Xmin_{bb(A)} > Xmax_{bb(B)}$
A southOf B	$Ymax_{bb(A)} < Ymin_{bb(B)}$
A above B	$Zmin_{bb(A)} > Zmax_{bb(B)}$
A below B	$Zmax_{bb(A)} < Zmin_{bb(B)}$

Table 4.10 Geometric specification of the basic direction relationships

1. Let the position of the ce-complex A be in the centre of the cube, i.e. in the cell (m_x, m_y, m_z).
2. $bb(A)$ stands for the "bounding box of e-complex A". The reader may imagine X and Y as horizontal and vertical lines, respectively.

The geometric specification is based on the absolute coordinates of the bounding boxes of the ce-complexes. Thus we have "broken down" the direction relationships to the Euclidean space. In some applications, however, it is of interest to define direction relationships, which are dependent on the position of the observer. For this we introduce the term "line of vision" which gives the direction, an observer is looking, if he fixes the direction relationships. The line of vision corresponds to the fixing of the coordinate system of an observer according to the coordinates of the refering system. As an example we give the definition of the relationships *"leftOf"* of two ce-complexes A and B (see fig. 4.36).

B leftOf A :=	(B nw A) $\vee$ (B w A) $\vee$ (B sw A),	if line of vision is N;
	(B w A) $\vee$ (B sw A) $\vee$ (B s A),	if line of vision is NW;
	(B sw A) $\vee$ (B s A) $\vee$ (B se A),	if line of vision is W;
	(B s A) $\vee$ (B se A) $\vee$ (B e A),	if line of vision is SW;
	(B se A) $\vee$ (B e A) $\vee$ (B ne A),	if line of vision is S;
	(B e A) $\vee$ (B ne A) $\vee$ (B n A),	if line of vision is SE;
	(B ne A) $\vee$ (B n A) $\vee$ (B nw A),	if line of vision is E;
	(B n A) $\vee$ (B nw A) $\vee$ (B w A),	if line of vision is NE.

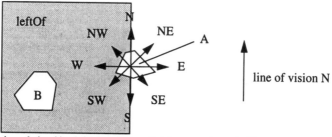

Fig. 4.36 Direction relationship "leftOf" with line of vision towards north (N)

The other direction relationships, such as *"rightOf"*, *"inFrontOf"* and *"behind"* can be defined correspondingly.

4.7.3 Topological Operators

a) Binary operators on connected e-complexes[1]:
   ECom × ECom → {ECom}

The topological operators can be specified in the same way as the topological relationships by the intersection and the union of the *boundary* @ and the *interior* $^{\circ}$ of connected e-complexes. Let A and B be two connected and convex e-complexes, then the *intersection,* the *union* and the *difference* of A and B are defined as follows[2]:

$$(\text{binOverlapOp A B}) := (@A \cap_{op} @B) \cup_{op} (@A \cap_{op} {}^{\circ}B)$$
$$\cup_{op} (@B \cap_{op} {}^{\circ}A) \cup_{op} ({}^{\circ}A \cap_{op} {}^{\circ}B)$$

1. In the following "ECom" stands for a connected e-complex.
2. $\cap_{op}$ stands for *binOverlapOp*, $\cup_{op}$ for *binUnionOp* and $\backslash_{op}$ for *binDiffOp*.

$$(\text{binUnionOp A B}) \quad := (@A \cup_{op} @B) \cup_{op} (^{o}A \cup_{op} {}^{o}B)$$
$$\setminus_{op} ((@A \cap_{op} {}^{o}B) \cup_{op} (@B \cap_{op} {}^{o}A) \cup_{op} (^{o}A \cap_{op} {}^{o}B))$$

$$(\text{binDiffOp A B}) \qquad := ((@A \setminus_{op} {}^{o}B) \cup_{op} (^{o}A \setminus_{op} {}^{o}B))$$

Note that the definition of the topological operators is valid for e-complexes of arbitrary dimension. The interior topologies which are generated during the execution of the operators are the union of the operator results between the single simplices. An example is the union of the results of all triangle intersections. Thus mostly a new triangulation of the result is necessary.

Let d be the dimension of the e-complex ($2 \leq d \leq 3$). We introduce the two auxiliary operators *cut* and *compose*, which are important for GIS applications. The *binCut-operator* cuts an e-complex along an $e_{(d-1)}$-complex (*"cutLine"* in 2D space, *"cutSurface"* in 3D space) and the *binCompose-operator* composes the two patches again.

$$\text{binCutOp}: \qquad E_{(d-1)}\text{Com} \times E_d\text{Com} \rightarrow E_d\text{Com} \times E_d\text{Com}$$

The *binCut*-operator cuts an e-complex at the given *cutline* and *cutSurface*, respectively. It returns the hereby generated two e_d-complexes.

$$\text{binComposeOp}: \quad E_{(d-1)}\text{Com} \times E_d\text{Com} \times E_d\text{Com} \rightarrow E_d\text{Com}$$

The *binCompose*-operator composes two e_d-complexes, which before were cutted along a *cutLine* (2D space) and a *cutSurface* (3D space), i.e. an $e_{(d-1)}$-complex. Let us now extend the topological operators for general not-connected e-complexes.

b) Topological operators on not-connected e-complexes:
$\text{ECom} \times \text{ECom} \rightarrow \text{ECom}$

The *intersection*, the *union* and the *difference* of two not-connected e-complexes A and B are defined as follows. Let $A_i \in A$, $B_j \in B$ be connected e-complexes.

$$(\text{overlapOp A B}) \quad := \cup_{op} (\text{binOverlapOp } A_i \ B_j)$$

$$(\text{unionOp} \quad A B) \quad := \cup_{op} (\text{binUnionOp } A_i \ B_j)$$

$$(\text{diffOp} \qquad A B) \quad := \cup_{op} (\text{binDiffOp } A_i \ B_j)$$

The new generated interior topology of the *overlap*-operator results from the union of the topologies of the binary executed intersections on the connected e-complexes A_i and B_j.

4.7.4 Metrical Operators

Distance measures (binDistanceOp : ECom × ECom → real)

As user-defined distance measures (metrics) on connected e-complexes we introduce the *minimal distance,* the *maximal minimal distance,* the *maximal distance,* the *distance between centroids* and the *Hausdorff*[1]*-distance* (GRÜNBAUM 1967, see. fig. 4.37).

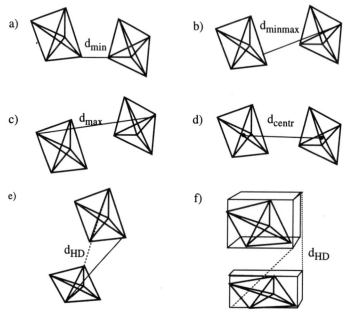

Fig. 4.37 User-defined distance measures on (connected) e-complexes:
   a) Minimal distance
   b) Minimal maximal distance
   c) Maximal distance
   d) Distance on centroids
   e) Hausdorff-distance
   f) Hausdorff-distance on bounding boxes

Let A and B be connected e-complexes, $a \in @A$ and $b \in @B$ be a set of all simplices on the boundaries of A and B. Then the introduced distance measures are defined as follows:

   a) d_{min} A, B := min (d(a,b))

   b) d_{minmax} A, B := max (sup d(a,b))

   c) d_{max} A, B := max (d(a,b))

   d) d_{Zentr} A, B := d(centr (A), centr (B))

---

1. Hausdorff, Felix (8.11.1868 - 26.1.1942), mainly worked in Bonn.

e) d_{HD} A, B := max (sup d(a, B), sup d(A, b))

f) d_{HD} bb(A), bb(B) := max (sup d(a, bb(B)), sup d(bb(A), b))

In $\mathfrak{R}^n$, d is equal to the Euclidean distance. We now try to estimate the quality of the different distance measures on e-complexes.

Quality of a distance measure (metrics) for spatially expanded objects in 3D space:

Let M be a set, $\mathfrak{R}$ the real numbers and p1, p2 two points in the Euclidean space. By a metrics we understand a mapping

d: M × M → $\mathfrak{R}$
   (p_1, p_2) → d (p_1, p_2)

with the following properties:

(a)   d (p_1, p_2) = 0 ⇔ $p_1 = p_2$ (identity)
(b)   d (p_1, p_2) = d (p_2, p_1) (symmetry)
(c)   d (p_1, p_3) ≤ d (p_1, p_2) + d (p_2, p_3) (triangle inequation)

If we extend d for distances between e-complexes with spatial expansion (dimension > 0), then for the *minimal distance* and the *distance on centroids* the *identity* (criterion a) is no more valid (see fig. 4.38). For the other distance measures, the distance between non-point objects cannot be zero. For the *minimal maximal distance* also the *symmetry* (criterion b) is no more valid. Thus the properties for distance measures only partially hold for e-complexes of arbitrary dimension. That is why we introduce the *"sensitivity"* of the distance measure refering the spatial expansion of the objects as an additional *quality measure* for distances between e-complexes.

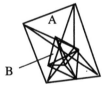

Fig. 4.38 Example for the anti-reflexivity of the minimal distance:
d_{min} (A, B) = 0 ⇏ (A = B)

The *distance with centroids* does not show any sensitivity refering the spatial expansion of the objects. The distance does not change with an expansion of the objects, even if overlappings occur, as we can see in fig. 4.37 d). On the *minimal distance* (fig. 4.37 a) the distance changes, if the objects are expanded "on their insides". Against this, on the *minimal maximal distance* the distance only changes, if the first object (left in fig. 4.37 b) is changed "on the inside" or if the second object (right in fig. 4.37 b) is changed "on the outside". Thus we receive different values for the distances with the change of the objects' order, i.e. the *minimal maximal distance* hurts the symmetric property of a distance measure. On the *maximal distance* (fig. 4.37 c), however, the distance only changes, if the objects are expanded "on the outsides". As the only of the introduced distance measures, the *Hausdorff-distance* considers the expansion of both objects into all directions. This means that for the determination of the distance, different sizes of the objects are taken into account in any case. A distance measure appears "natural", if it fulfils the intuitive expectations of the spatial terms "near" and "far away", like the Hausdorff-distance does. Thus we call the Hausdorff-distance a *"natural distance measure.* If we determine the distance between the convex hulls of the not-connected e-complexes, the introduced distance measures can be extended for general, not-connected e-complexes.

4.7.5 Geometric Properties

A *geometric property (geometric attribute)* gives a measue refering the geometry of an e-complex, such as its length, its surface or its volume. Let A be an e-complex of dimension d $(2 \leq d \leq 3)$.

geomAttribute : ECom $\rightarrow$ real

The geometric properties on e-complexes are defined on their simplices:

length (A)   := max $(d (a_i, b_j))$, with $a_i, b_j \in$ 0-simplices of A (points);

perimeter (A)   := $d(a_n, a_0) + \Sigma d(a_i, a_{i+1})$, with $a_i \in$ 0-simplices of @A (boundary points);

surface (A)   := Σ surface (a_i), with $a_i \in$ 2-simplices of A (triangles) for E_2Com and of @A for E_3Com, respectively (i.e. the surface of A);

volume (A)   := Σ volume (v_i), with $v_i \in$ 3-simplices of A (tetrahedra);

centre (A)   := centre of the boundary polygon and of the surface of A for E_2Com and E_3Com, respectively;

convexHull (A) := convexHull $(\cup a_i)$, with $a_i \in$ 0-simplices of A.

length (A) provides the largest spatial expansion of the e-complex A, i.e. the maximal distance between two 0-simplices of A. Self-explanatory, *perimeter(A)* provides the perimeter and *surface(A)* returns the surface of an e-complex[1]. *volume(A)* computes the volume of A as the sum of the volumes of its tetrahedra. By *centre(A)*, the gravity point of an e-complex is determined and *convexHull(A)* gives the convex hull determined by the hull of A's 0-simplices.

4.7.6 Geometric Cut- and Paste-Operators and Local Operations

Finally we introduce two auxiliary operators. With the *ClipBoxOp* a well-defined sub-e-complex (SECom) can be "cutted out" (see fig. 4.39). With the *setInBoxOp* this SECom can be "putted in" again, after a geometric update, i.e. the change of its simplices. As marginal condition we demand, however, that the boundary of the sub-e-complex (SECom)to be cut-

---

1. Volume (A) is only defined on e_3-complexes.

ted out, must not change. Thus only such sub-e-complexes can be cutted out, whose boundary is completely inside the box. If several boxes shall be cutted out from an e-complex, then we demand that they must not overlap.

ClipBoxOp   :   ECom   ×   Box   →   SECom

SetInBoxOp   :   SECom ×   ECom × Box   →   SECom

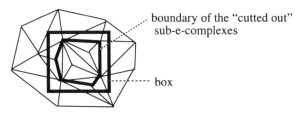

boundary of the "cutted out" sub-e-complexes

box

Fig. 4.39 "Cutting out" of a sub-e-complex from an e-complex inside a box

Obviously, the equal- and the disjoint-relationship can also be defined on sub-e-complexes of a single e-complex. Other topological relationships are only defined, if their result refering to the touching or the intersection of the e-complexes consists of complete simplices. I.e. there must not be sub-objects which are "smaller than a simplex". Thus all touchings and intersections can be topologically determined by searching the relevant simplices in the involved sub-e-complexes. That is why we can do without the specification of the boundaries and the interiors of the simplices[1] for all local topological relationships. We introduced this way of specification for global operations before. *Distance measures,* for instance, can be defined by means of the minimal distance between sub-e-complexes of arbitrary dimension. The *direction* between two sub-e-complexes can be determined with the method described in chapter 4.7.2.

With the *e-complex* we have introduced a spatial representation which is particularly suited for the representation and the management of the geometry and the topology of geo-objects. This is especially true for complex surfaces and volumes in 3D space. The explicit exploitation of the topology and the generation of *convex* volumes made possible the development of efficient geometric algorithms on e-complexes. With the ECOM-algebra, for the first time topological relationships were specified on the internal topology of the geo-objects. Furthermore, different classes of spatial operations were defined in 3D space. In the following chapters we demonstrate the practical use of e-complexes for the representation, management and the processing of geo-objects in a Geo-Information System. Particularly the *extensibility* of the e-complex to non spatial attributes (see chapter 5) has a favourable effect. Starting from the 0-simplices of the extended, so called "GEO-complexes" interpolations of arbitrary attributes on surfaces (triangle networks) or on the volumes (tetrahedron networks) can be computed. In chapter 6 we will see how the *decomposition* of the complex into simplices has

---

1. Such a simplification would even be possible for global operations, if we decomposed the e-complexes along their intersection borders during the insertion into the database. Thus the decomposed objects would be already stored during the insertion. However, this would lead to a significant increase of the object set.

a favourable effect on the running behaviour of geometric algorithms. In chapter 7 it is demonstrated, how the e-complex can be embedded into an *object model* of geologically defined geometries. Starting from the special requirements of the interactive 3D modeling and the spatial data management in the geosciences, we demonstrate by a geological application that the e-complex is particularly suitable for an efficient execution of *geometric and topological queries* on geo-objects.

Chapter 5

Integration of Building Blocks for e-Complexes into a GEO-Model Kernel

We designate a *GEO-model kernel* as a toolkit that supports the modeling and management of geo-objects for GISs similar to the way a CASE-tool supports the software engineering process. Essentially, the GEO-model kernel has to perform two tasks: The first is to support the design of extensible, open GIS tools like the geo-object modeling, the 3D visualization or the automatic checking of the integrity constraints. The second task is to provide base functionality for the management of geo-objects being represented as e-complexes. According to these tasks we subdivide the building blocks of the GEO-model kernel into two groups:

1) CASE-building blocks,
2) service-building blocks.

The first group of building blocks provides for example tools for the *design of object models*, the *specification of geometric datatypes* or the *checking of integrity constraints* (see fig. 5.8). Examples for the second group of building blocks that provide services to the geoscientific user, are a *repository* for the management of meta-data and services of the GEO-model kernel, *spatial access methods, (3D-) visualization* or the *transformation between different spatial representations*.

5.1 CASE-Building Blocks

An important CASE-building block of the GEO-model kernel is the support of the specification of abstract data types (ADTs). We will illustrate this with the example of an ADT "GEO-Complex" and the intersection of GEO-complexes. For this we extend the e-complex with thematic attributes and methods, model it as an ADT and call it "GEO-complex" from now on. Fig. 5.1 shows the abstract description of a GEO-complex.

5.1.1 Integration of Thematic and Spatial Attributes

The traditional proceeding for the attaching of thematic attributes to geo-objects is to assign one or more thematic attributes to a whole geo-object. But with the use of e-complexes the additional possibility exists to assign thematic attributes to the topological "substructures" of the e-complex, i.e. to its internal structures. For example internal neighbouring sets of edges, triangles or tetrahedra can be attached with thematic attributes by means of their topologies.

Extension of the e-complex with thematic attributes

Informally, a GEO-complex is an e-complex, i.e. a simplicial complex with exlicit topology (neighbourhoods) and geometry (set of 3D coordinates) extended with an Id (denotation), thematic attributes[1] and methods that are attached to the 0-, 1-, 2- or 3-simplices (see fig. 5.1).

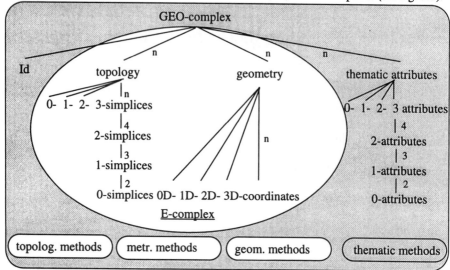

Fig. 5.1 ADT „GEO-Complex"

With this the possibility exists to assign the same values of an attribute to parts of an e-complex, i.e. to a set of simplices. Thus a grouping of attribute values by means of the topology is provided.

A subject that allows the clear demonstration of the attachment of thematic attributes to e-complexes in 3D-GIS is geology. Geology is one of the most obvious subjects for 3D-Geo-Information Systems, as it considers three dimensional structures and substance distributions that are developing in time. That is why the "world" of the geologist can only be caught incompletely without a threedimensional reference framework.

An application from geology

We will now demonstrate the integration of thematic and spatial attributes in 3D-GIS by the example of geologically defined points, lines, surfaces and volumes. Those can for example be points in space or digitalized lines from drillings, maps, sections or the seismics. Typical examples of geologically defined surfaces and volumes are boundary surfaces and volumes bounded by fault and layer surfaces, respectively.

---

1. An alternative approach is to consider the measure points in geology as information points and to interpolate those with the help of functions upon surface- and volume-attributes instead of attaching the thematic attributes to the 0-simplices.

a) Geologically defined points and lines

Pointlike geometries appear in geology for examples as *measure points* along *drillings* (see fig. 5.2). The specific attributes of the drillings like the number of layers, stratigraphy, lithology etc. can be represented as thematic attributes of 0-simplices (points). With special methods like kriging the punctually given thematic attributes can be interpolated upon the surface. Thus a more detailed placing of the thematic attributes is possible which is adapted to the patches of a surface. An example for the importance of lines (e_1-simplices) in geology are the layer and fault lines that result from the correlation of single drillings in section planes. Within the set of given intersections through a geological strata model the layer and fault lines can be represented as 1-simplices of an e_2-complex (see next paragraph). Typical attributes of a layer line would be the name of the attached layer, the lithology or the stratigraphy.

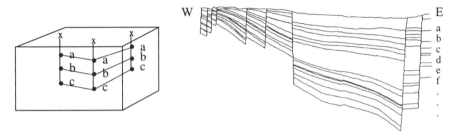

Fig. 5.2 Examples for geologically defined points and lines:
Borehole points (sketch) and section through a strata sequence with attached attributes
(a, b, c, ...) respectively

b) Geologically defined surfaces

In geology e_2-complexes are especially suited to model geological strata models. An example of a geological strata model shows fig. 5.3.

Fig. 5.3 Example of a geological strata model consisting of layer and fault surfaces
(from: ALMS et al. 1994)[1]

---

1. The picture was produced with the 3D-visualization tool GRAPE (1993), see also chapter 7.

For the user the possibility exists to consider the thematic attributes of the 0-, 1- and 2-simplices of the e_2-complex individually towards their "resolution". The most obvious way is to attach the thematic attributes to the 0-simplices (points). However, with interpolation functions on lines or surfaces (with supporting points) thematic attributes can also be attached to 1-simplices (edges) and to 2-simplices (triangles). Thus for example lines and surfaces with the same thematic attribute *colour* can be represented in the same colour.

c) Geologically defined volumes

The insight of the spatial relationships between the participating volumes is permanently improved by the interactive reconstruction process of geologically defined 3D geometries executed by the geologist. Thus the existing knowledge about the history of the geological development may be corrected (SIEHL et al. 1992). With the change from the e_2-complexes to the e_3-complexes typical attributes like stratigraphy, lithology etc. may be attached to geological layers in a volume model. This has the advantage that the volume of the layers, i.e. the *internal topology,* may be also attached with thematic attributes.

<u>*Intersection of GEO-complexes*</u>

It is an open question how the thematic attributes of GEO-complexes are changing, if geometric operations are applied to the GEO-complexes. We explain the problem by the example of the "intersection" of two independently determined aquifers A and B that are drilled by wells (measure groups of A and B). Let A and B be represented by GEO-complexes. As thematic attributes, we pick out the percentage of sand and the porosity. For water supply it is interesting to determine the maximal hauling of water. One possibility could be to "intersect" A and B by taking the union of the points of A and B which are lying inside the surface of C, that A and B both overlap. Additionally, at the intersection points of the triangulations of A and B "virtual measure points" are generated. To those virtual points new values may be assigned with an interpolation function (see the white points in fig. 5.4a), whose points are dependent on the value of the new neighbouring measure points. Thereby a new distribution for the values controlling water supply is computed for the common surface, i.e. a new triangulation is necessary. With an interpolation function the data controlling water supply can be interpolated upon the surface with the real and the "virtual" measure points as input.

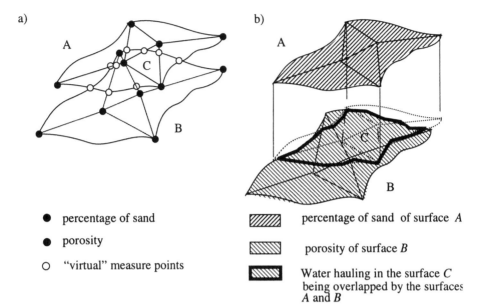

● percentage of sand	▨ percentage of sand of surface A
● porosity	▨ porosity of surface B
○ "virtual" measure points	▨ Water hauling in the surface C being overlapped by the surfaces A and B

Fig. 5.4 Intersection of two GEO-complexes with the thematic attribute "lithology"
a) dynamic computation of the thematic attribute values on points
b) static overlay of surfaces

A second possibility is, similiar to the layer technique of today's GISs, to start from the distributions of the percentage of sand and the porosity. Thus for every triangle[1] being generated during the intersection, the union of the values for both distributions for the respective face of A and B has to be computed by an overlay of A and B. Therefore a new interpolation is not necessary, i.e. the newly supplied values for the water can directly be "picked off". Whereas the first possibility computes the values *dynamically,* with the second possibility for every triangle a union of existing, i.e. *static* attribute values is effected. For both methods additionally the possibility of an *interactive intervention* by the geoscientific expert should be provided.

5.1.2 Object Oriented Modeling Support and Integrity Checking

The graphical support for the design of object models is a fundamental functionality for the GEO-model kernel. With an editor that provides a toolbox-like composing of single primitives for object classes and relatinships, for instance, the geoscientific expert can be efficiently supported during the design of object models with an extended Object Modeling Technique (RUMBAUGH et al. 1991). Integrity constraints for geo-objects can also be integrated into the object model. Automatically observed spatial integrity constraints, for example, can check the plausibility of spatial relationships between geo-objects. We distinguish between the following *classes of spatial integrity constraints* for the checking of

---

1. As the intersection of two triangles can also result in four-, five- or six-gons, additional triangulations may be necessary.

- local and global topological relationships (neighbourhoods etc.),

- metrical relationships,

- direction relationships,

- the description of the geometry of a geo-object,

- the membership of a geometry (point, line, surface etc.) to a geo-object.

Examples for the checking of a *topological relationship* as an integrity constraint in a typical geological application, which is based upon the law of superposition, are: "Stratigraphic surfaces must not cross each other" or "as *B* is older than *A, A* always is the hanging stratum of *B*". *Metrical relationships* are playing a dominant role for the checking of the distance between different faults. The geologist knows for example of a certain fault that it is situated north of another one. This can be tested by the checking of *direction relationships* between the faults. The *description of the geometries* can be used for the checking of the stratigraphic sequence and the domain of permissible depths of layers or for the testing of a minimal volume of a layer (for example the volume must not be negative). Finally, an important clue for an eventually incorrect digitalization is the checking of the geometry, i.e. that points, lines or triangles of a surface must occur only once per surface.

As we will see in the following, the number of values for the thematic values of the simplices of an e-complex may be significantly reduced with the help of integrity constraints. Thus with the goal to reduce the used storage space we can distinguish between simplices "with and without thematic information". We first give some examples for topological integrity constraints[1] that support the *reduction of storage space* and can be seen as part of a building block for the checking of the integrity of GEO-complexes:

- A value of an thematic attribute is attached to a *tetrahedron (3-simplex)*, if and only if its four neighbouring tetrahedra have different values of this thematic attribute,

- A value of an thematic attribute is attached to a *triangle (2-simplex)*, if and only if its three neighbouring triangles have different values of this thematic attribute.

Both integrity constraints effect that values for the thematic attributes of the 2- and 3-simplices are only given, if they are situated at the boundary of a set of simplices with equal attribute values. I.e. values are only given for simplices that are situated at the border line of a surface and volume. Therefore, relative to a given thematic attribute, we can distinguish between boundary tetrahedra and boundary triangles and interior tetrahedra and triangles, respectively. Thus thematic information is only attached to the boundary tetrahedra and triangles.

---

1. If the values of the thematic attributes were computed by functions over information points (interpolation), the dependencies of the integrity constraints would not only consist of the neighbouring simplices, but also of the interpolation functions.

5.2 Service-Building Blocks

As we already treated the transformation of different spatial representations in chapter 4.7, and as we will enter into the 3D-visualization in connection with the interactive query support in chapter 7, we will not illustrate these two service-building blocks in detail. Therefore we immediately proceed to the access building block for GEO-complexes.

5.2.1 Access Methods for GEO-complexes

One of the most important service-building blocks of the GEO-model kernel has the task to provide an efficient access on GEO-complexes in secondary storage. Such access methods for GEO-complexes have to support three kinds of query types:

    - queries on thematic attributes,
    - queries on the geometry and topology of a set of GEO-complexes,
    - queries on specified parts of the geometry and topology of a single GEO-complex[1].

If the queries on thematic attributes predominate, the objects should be *clustered* towards their class memberships and indexes should be generated on the thematic attributes. But if almost only spatial queries occur, it is ingenious to store the GEO-complexes clustered towards their spatial position.

1) <u>Queries on thematic attributes</u>

For queries on thematic attributes of GEO-complexes we can fall back upon standard access methods like the B-Tree (BAYER and McCREIGHT 1992). For combined queries, i.e. queries upon thematic *and* spatial attributes, either separated access methods can be applied or the queries on the thematic attributes must be treated as zero-dimensional queries in the spatial access method. These are the so called point queries.

2) <u>Queries on the geometry and topology of a set of GEO-complexes</u>

For the support of the spatial search[2] spatial access methods were developed since the middle of the 80ties. They first were only designed for the access on points (NIEVERGELT and HINTERBERGER 1984) and rectangles (GUTTMAN 1984) in 2D space. They are, however, extensible for arbitrary objects in 3D space. In the last years numerous optimizations of spatial access methods were proposed like the R^+-Tree (SELLIS et al. 1987), the Cell-Tree (GUNTHER 1989), the R*-Tree (BECKMANN et al. 1990), the Buddy-Tree (SEEGER and KRIEGEL 1990) or the TR*-Baum (SCHNEIDER 1993), to name only a few. However, none of these methods could decesively assert itself against the other ones. This was also confirmed by later results (GAVRILA 1994). Today we can state that altogether the access methods behave similarly. Their suitability is dependent on the kind of data as well as on the query mix. Hereby the *degree for the overlapping of the data* and the *average size of the*

---

1. For example the search of starting triangles for intersection algorithms (see chapter 4.5).
2. A typical query with a spatial search is: "Provide all objects that are inside a specified box or that intersect the box".

query space play a special role. WATERFELD (1991) has found with the DASDBS Geokernel (WATERFELD and BREUNIG 1992) by means of cartographic data of a Germany
map that the R-Tree, a *dynamic access method* [1]is optimal for the "complete query"[2] and that
it is tendentiously best suited for big query windows and high data overlappings. The "clipping" Grid File, a *static access method*[3], had an optimal runtime behaviour for point queries.
It was disadvantageous for high data overlappings, but partially optimal for small query windows and small overlappings. The *quadtree*[4] showed off to be a compromise between the
extreme cases of the R-Tree and the Grid File. The quadtree transforms arbitrary geometries
into squares in a preliminary step. The approximation of the geometries results from a regular
and hierarchical partitioning of space. The squares can be lexically sorted as quadtree codes,
i.e. variable long digit sequences of the base 4, in a B*-Tree. As the measurements of WA-
TERFELD (1991) showed, the quadtree was best suited for queries with a medium large query window and for larger data overlapping. Transferred to 3D space, we can expect by the
octree average runtime results for the access on complete GEO-complexes. An alternative
access method is the TR*-Tree (SCHNEIDER 1993) which stores decomposed geometries
in its leaf nodes and tendentiously behaves like the R-Tree. It gains a better distribution of
the geometries in the leaf nodes than the R-Tree by means of optimized split strategies.

Besides the search for complete GEO-complexes in space, the search for specified parts of
GEO-complexes plays a central role for the support of spatial operators.

3) Queries on specified parts of the geometry and topology of a single GEO-complex

Often it is of interest to access directly to specified parts of geometries and topologies, like
triangles or tetrahedra of a single GEO-complex. Such an access can also efficiently support
the search of the starting triangles during the intersection of two GEO-complexes (see algorithm *overlapOp*, step2, chapter 4.5). The intersection of two convex and concave e_2-com-
plexes is outlined in fig. 5.5 a) and 5.5 b), respectively.

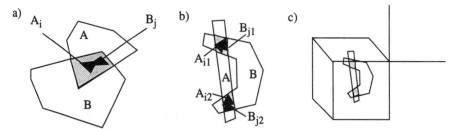

Fig. 5.5 a) Intersection of two convex GEO$_2$-complexes
     b) Intersection of two concave GEO$_2$-complexes
     c) Several connected patches in *one* cell of the access method

---

1. I.e. the paritioning of space is changing with the intersection of new data.
2. The result of the "complete query" is the sum of all objects in the complete space.
3. I.e. the partitioning of space is decided in advance. The clipping Grid File cuts the geometries in the
buckets, if they overlap the boundaries of the buckets.
4. Realized on top of the B*-Tree.

a) Intersection of two convex GEO₂-complexes

During the intersection of convex GEO_2-complexes an arbitrary pair of triangles $(A_i \in A, B_j \in B)$ has to be found so, that A_i and B_j are both situated in the intersection area of the bounding boxes of A and B and so that "A_i intersects B_j" is valid.

In chapter 2.2.4 we shortly introduced the PM-Quadtree (SAMET 1990a). In the PM-Quadtree the leaf nodes are distinguished depending on the "filling degree", i.e. if the node is completely filled ("black node"), partially filled ("grey node") or not at all filled ("white node"). Let from each leaf node point a reference to the real data, i.e. to the d-simplices of the GEO-complexes that are totally or partially inside the octree cell of the corresponding leaf node. Thus we can use an octree-variant of the PM-Quadtree for the search of the starting triangles of the intersection algorithm of two GEO-complexes in the following way:

algorithm *triangleSearch*
{Let A and B be two convex GEO-complexes of dimension d $(2 \leq d \leq 3)$ in 3D space. Search for a pair of d-simplices $(A_i \in A, B_j \in B)$ so that "$A_i \cap B_j$" provides the value *true*}

Step 1:   Intersect the octree codes of both GEO-complexes A und B;

Step 2:   if a leaf node exists that after the intersection is a black node:
              (a) Intersect in the leaf node a d-simplex of A with a d-simplex of B (with the
                  circle method) until the intersection provides the value TRUE or until no
                  more d-simplices are existing. Mark the both d-simplices A_i and B_j as *star-
                  ting simplices* (starting triangles and tetrahedra, respectively) for the inter-
                  section algorithm;
              otherwise:
                  (b) Do step 2a for the intersection of the black with the grey leaf nodes. If no
                      more starting simplices are found, do step 2a for all grey leaf nodes;
                  <u>return</u> A_i, B_j
end *triangleSearch.*

FERRUCCI and VANECEK (1991), BRUZZONE et al. (1993) and DE FLORIANI et al. (1994) have proposed a hierarchical representation of triangle networks for the modeling of digital terrain models. With such a representation the search of the starting triangles can be directly executed on the spatial representation, but the preselection of triangles that intersect each other is evidently more costly than with the octree.

b) Intersection of two concave e-complexes

For the intersection of two concave GEO-complexes A und B the access method must find a pair of d-simplices $(A_{ik} \in A, B_{jk} \in B)$ so that A_{ik} intersects B_{jk} (see fig. 5.5b) for each connected area that is generated with the intersection of both GEO_2-complexes. Thus the goal is to identify *all* of the connected areas in which intersections occur. In each of these areas a pair of d-simplices has to be searched for the start of the intersection algorithm. The octree, however, is not suited to determine all of the intersecting areas. For example several connected areas can be inside a single octree cell of the highest resolution, i.e. the areas do not distinguish themselves in their octree codes (see fig. 5.5c). The octree access method can only be used to determine the intersection of the whole octree codes of the both intersecting

GEO-complexes. This, however, does not lead to an advantage, because all the d-simplices of the first GEO-complex have to be tested for the intersection with all of the d-simplices of the second GEO-complex. Thus the generation of convex e-complexes is advantageous in any case.

5.2.2 Spatial Relationship Graphs

It is convenient to store precomputed spatial relationships between GEO-complexes, if they are often needed in an application. Therefore our second service building block serves for the explicit storing of the results of spatial queries. For line networks in 2D space the storage of spatial relationships in a database by means of graphs was already proposed by ERWIG and GÜTING (1991) and GÜTING (1991), respectively. *Constraint-description graphs* (ZALIK et al. 1992) serve for the description of relationships between parts of the geometry of a single object. An example is the relationship "right angle" between two lines belonging to a single object. However, it is not our goal to store spatial relationships between the simplices of a single GEO-complex in a graph. The topological relationships between single simplices of an GEO-complex are many to store them explicitly. The point is the precomputing of spatial relationships between different GEO-complexes. Examples are the distance relationship between geological layers or the direction relationships between geological faults. In the following example (fig. 5.6) the direction relationships shall be described between the four geological faults *A, B, C* and *D*. The nodes of the graphs describe the geo-objects (GEO-complexes) and the names of the edges stand for the spatial relationships between them. The complete relationship table[1] is shown in table 5.1. For direction relationships it is sufficient to consider the table above the diagonal. The inverse relationships (for example *EastOf* for *WestOf*) can be stored in advance, respectively.

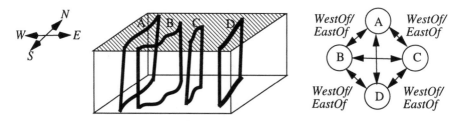

Fig. 5.6 Example situation for the position of geological faults with their relationship graph

Note that our description for spatial relationships, the graph, may again be considered as a 1-complex. Consequently, with the GEO-complex we have found a unified description for geo-objects and spatial relationships between the geo-objects.

---

1. The graph is to be read in direction from the lines to the columns, because the order of the execution is decisive for direction relationships.

→	A	B	C	D
A	-	WestOf	WestOf	WestOf
B	EastOf	-	WestOf	WestOf
C	EastOf	EastOf	-	WestOf
D	EastOf	EastOf	EastOf	-

Table 5.1 Relationship table of the situation of fig. 5.6

We may use graphs to describe spatial relationships independently of the coordinates of the GEO-complexes. BOUILLE (1976) has shown how one may transfer spatial relationships between the layers of a geological map into a graph (see fig. 5.7).

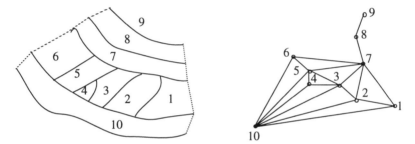

Fig. 5.7 Geological map and its representation in a graph (taken over from: BOUILLE 1976)

The geological map can be transferred into a graph with a simple determination of the neighbourhood relationships of the faces. The nodes of the graph again describe the geo-objects (geological layers) and the edges express the spatial relationships between them. The spatial relationships may be derived simply over the length of the paths between the nodes of the graph or with direct connections between the nodes. Therefore, the spatial relationships need not explicitly be named at the edges of the graph as in the example before. Two layers are *neighbouring,* if a direct connection between the two corresponding nodes in the graph exist. A layer A *lies above* a layer B, if the shortest path to A is shorter than the shortest path to B, starting from the starting node[1], respectively. A layer *lies directly above/below* another one, if its nodes in the graph are neighbours and if the length of their shortest paths from the starting node differ by one. Nodes in *the same depth* of the graph describe layers in the same depth.

---

1. The starting node in fig. 5.7 is the node with the number 9.

5.3 Integration of the Building Blocks into the GEO-Model Kernel

Our goal is to integrate the building blocks of the GEO-model kernel into an object oriented database management system (ODBMS). Today's commercially available "ODBMSs" are client-server systems, whereby the methods have to be implemented in the client. The server, however, serves for the pure data management, i.e. for the loading and the storing of data. Because of the missing functionality in the server for implementing methods, the building blocks of the GEO-model kernel have to be developed "on top" of the ODBMS. Against that with an extensible database kernel system like the DASDBS Geokernel (PAUL et. al. 1987; SCHEK and WATERFELD 1986; BREUNIG et al. 1990; WATERFELD and BREUNIG 1992) or OMS (BODE et al. 1992), user defined datatypes can be embedded deeply into the system architecture (see fig. 5.8). By that time critical building blocks like geometric operators or spatial access methods can be efficiently realized with a narrow coupling close to the internal storage representation of the database kernel system. The geometric operations to be introduced in chapter 6 are implemented in this way within the type manager of OMS. *Storage Object Cache* and *Storage Object Manager* provide the framework for the management of OMS objects in the main memory and the secondary storage, respectively. The type manager of OMS, in which a more complex structure is "coined" to the storage objects, can be compared with the Cluster- and *Access Manager (*AM) in the DASDBS Geokernel, which can load and store complex records of the *Complex Record Managers (CRM)* by a buffer (so called transfer areas) into and outof the main memory, respectively. The DASDBS Geokernel is a storage interface at which NF^2-relations can be defined and processed[1]. The *AM* is directly put on the DASDBS-kernel that provides functions for the management of complex objects by the *Complex Record Manager (CRM)*. The functions are mapped upon the *Stable Memory Manager (SMM),* a set-oriented interface based on pages. Fig. 5.8 shows the embedding of building blocks of the GEO-model kernel into the extensible database kernel systems DASDBS Geokernel and OMS.

---

1. The so called non-first-normal-form relations (JAESCHKE and SCHEK 1981) have the particularity that they may have relation-valued attributes.

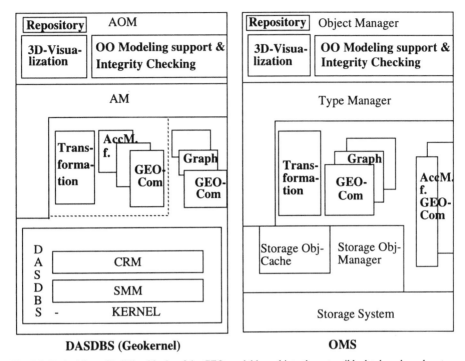

DASDBS (Geokernel)   **OMS**

Fig. 5.8 Embedding of building blocks of the GEO-model kernel into the extensible database kernel system DASDBS Geokernel and OMS

Whereas in the DASDBS Geokernel the GEO-complexes and the graphs have to be trans-formed as "EDTs" (**E**xternal **D**efined **T**ypes) into the internal database representation by the application programmer, OMS additionally provides the possibility to realize new types as "IETs" (**I**nternal **E**mbedded **T**ypes) directly in the storage object representation of OMS. On the one hand this deep embedding into the database kernel system may lead to a gain in ef-ficiency, for example for the execution of geometric operations (see NOACK 1993). On the other hand, data independency is hurted. In any case, the minimization of the transformation costs should be considered for an efficient support of the operations designing the mapping of the GEO-complexes upon the internal database representation of DASDBS and the stora-ge object representation of OMS, respectively. Therefore it is advantageous to map as big parts of one e-complex as possible into the internal database representation, for example its e_0-, e_1-, e_2- and e_3-simplices each, as a unit. Access methods *(AccM)* and the building block for the *transformation* of different spatial representations into the e-complex representation can both be embedded in an abstract interface that allows extensions for new implementati-ons at any time. In the DASDBS Geokernel additionally the spatial clustering of complex records is supported. In OMS access methods can directly use the storage object representa-tion. Building blocks for a *repository*, for the support of the *object oriented modeling, integrity checking* and the *3D-visualization, respectively* can be integrated into the *Application Object Manager (AOM)* of the geo-database of DASDBS and into the *Object Manager* of OMS.

Last, but not least, with regard to new GIS requirements such as complex modeling (for example the prediction of the grain harvest in a given area, environmental analysis and other multi-disciplinary applications (OLIVER and WEBSTER 1990; HACK and SIDES 1994; VAN OOSTEROM et al. 1994), *interfaces with external GIS-tools* become more and more important. Therefore an *open architecture* of the GEO-model kernel is of special importance. RAPER and RHIND (1990) are speaking of the development from *function oriented* to *task oriented* GIS. First approaches towards the direction of *interoperable GIS* start from a narrow coupling such as realizing a user interface on top of several GIS (RAPER and RHIND 1990) or implementing a communication by means of a file exchange (ABEL et al. 1992). In (BREUNIG and PERKHOFF 1992) an approach for the integration of heterogeneous GIS, a *Data Integration System*[1] *(DIS)*, is introduced. A further step is the integration of GIS data by means of an extensible DBMS (HAAS and CODY, 1991). A GEO-model kernel being embedded into an open DBMS could server as a mediator between GIS and geo-services. SCHEK and WOLF (1992) also speak of "autonomous operation services" on which the database shall access in cooperation with other software systems like GIS. A precondition is that the GEO-model kernel has knowledge about the architecture and the functioning of the operation services. Furthermore, it must have knowledge about a specification of the data handling component of each operation service. Furthermore, it must have a disposal of a specification of the data handling component of each operation service. WOLF et al. (1994) introduce a network based architecture for a GIS which provides a communication between external services. The integrated use of external services, however, raises many unsolved problems, such as the solution of semantic conflicts, *message passing across system borders* or the question, which functions of a GIS should be realized centrally and which should be solved locally as operation services. We cannot deepen these questions here, as they are extensive, separate research topics since several years (ACM 1990; WONHDBS 1990; WORBOYS and DEEN 1991; SCHEK and WOLF 1992; WOLF et al. 1994).

---

1. In the DIS query expressions of the different database base languages are to be translated into a functional language (PERKHOFF 1991).

Chapter 6

Performance Behaviour of Geometric and Topological Algorithms

Subject of this chapter is the evaluation of measurement series of geometric and topological operations on e-complexes in 3D space and their embedding into typical GIS queries on the base of an extensible database kernel system.

6.1 Goals of the Measurements

With the embedding of geometric and topological operations[1] into an extensible database kernel system as a possible implementation base for the GEO-model kernel, we can check the efficiency of the e-complex representation.

On the one hand, it is our goal to execute measurements[2] for spatial objects that are used in a real geoscientific application[3]. On the other hand, it is of interest to contrast the e-complexes with another spatial representation.

We subdivide the geometric objects used in the measurements into e_2- and "$e_{2\ 1/2}$"-complexes. An $e_{2\ 1/2}$-complex means a solid, whose surface is represented by triangulated areas (2-simplices) inclusively the explicit neighbourhoods of the 2-simplices.

6.2 Data Sources and Mapping of the e-Complexes upon an Extensible Database Kernel System

First we introduce a realization of e-complexes on the internal structures of an extensible database kernel system. For this we choose the Object Management System (BODE et al. 1992) that enables the user to realize user defined data types as so called Internally Embedded Types directly upon the internal structures of the StorageObjectManagEr. An OMS object may consist of hierarchically structured collections of untyped and variable long byte strings. The task of the SOME is to map the storage objects upon the secondary memory.

---

1. i.e. relationships and operators.
2. Most of the measurements were executed within the diploma thesis of NOACK (1993) and SCHOENENBORN (1993). The hardware platform was a SUN SPARC 10, model 20 (33MHz) under SunOS 4.1.3, which is a registered trademark of Sun Microsystems, Inc. As an implementation base the extensible database kernel system OMS (BODE et al. 1992) was used.
3. The GIS application will be exclusively treated in chapter 7.

An e_2-complex may be represented as an IET in a storage object list with the following seven atomic OMS storage objects which represent variable long byte strings, respectively:

e_2-complex (denotation, bounding box, 0-simplices, 1-simplices, 2-simplices,
number of 0-, 1-, 2-simplices, 3D-coordinates (x, y, z))

The *denotation* consists of the name of the objects and of the date it was generated etc. It uniquely identifies the e-complex. *0-, 1-, 2-simplices* stand for the topology of the 0-,1- and 2-simplices, i.e. each 2-simplex is represented with its 0-simplices and three references to its neighbouring 2-simplices. *number of 0-, 1-, 2-simplices* declares the number of simplices of the e_2-complex. Finally, *3D-coordinates* describes the geometry of the 0-simplices, i.e. their x-, y- and z-coordinates.

The storage object representation in OMS presented above enables the user to load the bounding box separately from the geometry and the topology of the e-complexes into main memory. Furthermore, the topology may be selected towards its dimension, i.e. 0-, 1- or 2-simplices may be treated separately. The storage object representation can be directly mapped upon an efficient main memory representation (SCHOENENBORN 1993). The OMS storage objects for 0-, 1- and 2-simplices can be implemented as arrays to enable a fast access. The operators were realized with the so called operator graphs (BODE et al. 1992) that allow the control of arbitrary function sequences in the database kernel system.

The measurements were executed basing on three different data sources. The geological data (e_2-complexes - data source 1) used in the GIS application were generated by means of plans, i.e. cross sections[1], which were digitalized profilschnitte at the Geological Institute of the University of Bonn. They are triangulated 3D areas of geological layers of the Lower Rhine Embayment in the area of the Erft block. Their size is varying between 12 and 124 triangles per e-complex (see table 6.1). To make suitable statements for queries, each of the 16 triangulated areas was duplicated 20 times so that totally 320 objects were examined as a data source.

Data source	# Objects	Kind of objects	Representation	# Areas	# Edges
1	320	areas	e_2-com	7 - 124	18 - 224
2a	1000	cuboids	Brep	6	12
2b	1000	cylinders	Brep	36	102
3a	1000	cuboids	$e_{2-1/2}$-com	12	18
3b	1000	cylinders	$e_{2-1/2}$-com	132	198

Table 6.1 Characteristic values of the different data sources

---

1. The cross sections were kindly provided by the RHEINBRAUN AG.

The further two data sources are 100 artificially generated solids (cuboids and approximated cylinders), respectively. They were generated in the vector representation[1] (boundary representation with polygons as base areas) as well as in a triangulated surface representation, i.e. as $e_{2\,1/2}$-complexes. The data sources 3a and 3b were directly generated from the sources 2a and 2b with an appropriate transformation (triangulation). The cylinders were generated with rotation and clipping between two cuboids. During the triangulation of the cuboids and cylinders the method of the biggest angle was used, i.e. proceeding from an edge of the polygon, the next triangle was connected with the point whose edge included the largest angle with the given edge. This proceeding avoids the generation of triangles with acute angles. To reach a likely "natural" spatial ordering, the objects were equally divided in space (equally divided randomized translation). Subsequently, the objects were subdued a randomized axis rotation. The cylinders were approximated with 36 faces. Table 6.1 contains the characteristic values of the data sources.

6.3 Spatial Operations and Queries

All of the examined spatial relationships and operators were embedded into OMS (BODE et al. 1992). Hereby a loose coupling[2] was realized, i.e. the e-complexes were loaded with big units[3] into the main memory for a further processing. We measured the needed real time in average (process and system time) of the operations over a sequence of 1000 calls. For three typical spatial queries the whole real time needed for the query was measured. All the measurements were executed without spatial access support.

As typical spatial operations the *inside-, outside-* and *overlap-relationship,* as well as the *overlap-operator* and the *distance-operator* were examined. For the $e_{2-1/2}$-complexes (triangulated cuboids - data source 3a) and for the cuboids represented in the boundary representation, all of the operations were measured. Additionally, for the *overlap-relationship* the cylinders (data source 3b) were added, to obtain a comparison with different complexities of objects. Six measurement series were executed (three for triangulated cuboids and triangulated cylinders, respectively). The test object was varied by a simple triangulated cuboid (72 triangles), a more complex triangulated cuboid (136 triangles) and a simple triangulated cylinder (132 triangles) (see also table 6.3).

The formulation of the queries was effected with the so called "base model language" of OMS. The SELTRANS-operator, which occurs in the following queries, is a general form of the selection and projection known from the Relational Algebra. The SELTRANS-operator applies a transformation to all elements of a set which fulfil a certain predicate. If the predicate is fulfilled, a copy of the result is taken over into a so called *ResultBag.*

Three queries were chosen that are typically for GIS: they contain a *distance-query,* a combined *inside- und outside query,* and *an intersection between GeoObjects* as query predicates:

---

1. The realized representation does not include topological information *(no winged-edge representation).* As geometric information, the point coordinates of the polygons and the plane equations of the solid surfaces were given.
2. The integration of internally embedded types with the help of the so called operator graphs is described by BODE et. al (1992) in detail.
3. Bounding box, geometry and topology are realized in one unit, respectively.

(Q1)   *SELTRANS [RANGE o: **distance** (o, cylinder) < radius,*
                      INSERT (ResultBag, o)] (GeoObjects)

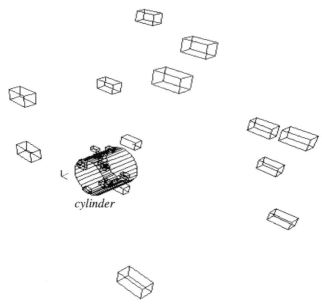

Fig. 6.1 Visualization of the query result from query Q1 with data source 2a

(Q2)   *SELTRANS [RANGE o : **inside** (o, cuboid1) AND **outside** (o, cuboid2),*
                      INSERT (ResultBag, o)] (GeoObjects);

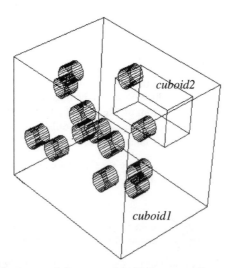

Fig. 6.2 Visualization of the query result from query Q2 with data source 2b

(Q3)   *SELTRANS [RANGE o : **overlap** (o, cuboid),*
                         INSERT (ResultBag, o)] (GeoObjects);

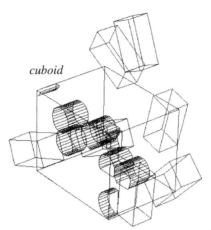

cuboid

Fig. 6.3 Visualization of the query result from query Q2 with data source 2a and 2b

Query *Q1* selects all *GeoObjects,* whose distance to the *cylinder* is smaller than the value of *radius*. Against that, query *Q2* selects all *GeoObjects* that are totally contained in *cuboid1,* but that are outside of *cuboid2*. Finally query *Q3* selects all *GeoObjects* whose intersection with the specified *cuboid* is not empty.

6.4 Results of the Measurements

6.4.1 Spatial Operations

Table 6.2 shows the results of the single operations for the triangulated areas (data source 1).

Operation	Data source	Kind of objects	t_{op} per hit	Reduction bb-test [%]
inside	1	Faces	120.83	45
outside	1	Faces	135.83	50
overlap	1	Faces	165.27	50
overlapOp	1	Faces	1190.14	50
distanceOp	1	Faces	9.52	-

Table 6.2 Measurement results of spatial operations for triangulated areas in ms

t_{op} *per hit* shows the average real time of the spatial operations for the case that a hit took place, i.e. if the query predicate was true. With exception of the *overlapOp-algorithm* the times per hit run up to about 100ms. *Reduction bb-test in* % shows how many percent of triangles could be spared with the spatial preselection, i.e. the *bounding-box test*. Thus this parameter shows the degree of the spatial overlapping of the objects[1]. In the algorithm only those triangles are considered that are inside the intersection of the bounding boxes of the two e-complexes to be compared. In average 45% - 50% of the examined triangles could be eliminated with the bounding-box test for all topological algorithms. Although in the *inside-algorithm*[2] more triangles have to be examined with a sequential scan, it came off better than the *outside-* and the *overlap-algorithm*. This can be explained with the use of the *circle method* introduced in chapter 4. Thus the neighbouring triangles were examined circle by circle. The relatively small reduction of triangles of only 35% in comparison with the 50% for the other operations are due to the fact that in the *inside-algorithm* only the number of triangles of the exterior object can be reduced. In the other algorithms the reduction consists of the spared triangles of both objects.

For the *overlapOp-algorithm*, which computes the intersecting geometry of two e-complexes inclusively the computation of a new triangulation, an algorithm was used that runs in quadratic time towards the number of triangles in the intersection area of both bounding boxes. Responsible for the high running time of more than one second is the fact that inside of this operator a result object has to be generated in OMS (see also SCHOENENBORN 1993). However, the performance could be tuned, if the topology was exploited in a way similiar to the inside-algorithm. The *distance-operator* has a constant runtime of 9.51 ms in average. The time is independent of the representation of the objects as the distance between two objects is always computed with the distance of the bounding boxes. Table 6.3 shows the results of the single operations for the triangulated cuboids and cylinders (data sources 3a and 3b). However, the results cannot directly be compared with the results of the triangulated areas. A spatial operation of 2D objects in 3D space cannot be extended without further ado to an operation of 3D objects in 3D space. The *inside-relationship*, for example, in the first case is synonymous to the "touching" of all parts of the objects, but the *inside-relationship* for two solids means that the volume of the first object is totally inside the volume of the second object. The other considered operations, however, are comparable. As result we can learn from table 6.2 and 6.3 that altogether the measured times for simple triangulated solids (cuboids and cylinders) are 15 - 20% faster than those of the triangulated areas.

Operation	Data source	Kind of objects	t_{op} per hit [ms]	Reduction bb-test [%]
inside	3a	cuboids	101.68	30
outside	3a	cuboids	115.26	50
overlap	3a	cuboids	114.08	50
overlap	3b	cylinders	149.21	40
distanceOp	3a	cuboids	9.51	-

Table 6.3 Measurement results of spatial operations for triangulated cuboids and cylinders in ms

1. The overlapping is reversely proportional with the number of the spared triangles (reduction).
2. As we defined in chapter 4.5, the inside-algorithm on two e_2-complexes in 3D space delivers the value *true*, if one triangle network completely touches the other one.

This can be explained by the small number of triangles (12) for the cuboids in comparison with a number of up to 132 for the triangulated areas. For cylinders (consisting of 132 triangles) the algorithm is about 50% more slowly than its pendant for 2D objects. In average, with 40% reduction of the triangles during the bounding-box test still 53 triangles (40% of 132) per e-complex had to be further examined in the *overlap-algorithm* for the cuboids in comparison to only 6 triangles (50% reduction) for the not triangulated cuboids. For the *inside-algorithm* in average only 30% of the triangles could be spared, because the access on triangles could be only reduced for one object, as we explained above.

6.4.2 Spatial Queries

Table 6.4 shows the measurement results of the spatial queries of chapter 6.3 with the triangulated cuboids (data source 3a). As we expected, the distance query (Q1) came off with the best runtime with 9.51ms per hit.

Query	Data source	Kind of objects	Hit rate [%]	t_{op} [ms]	t_{op} per hit [ms]	Reduction [%]
Q1	3a	cuboids	100	9.51	9.51	-
Q2	3a	cuboids	11	29.67	192.72	50
Q3	3a	cuboids	9	23.19	167.16	80

Table 6.4 Measurement results of typical spatial queries for triangulated cuboids in ms

For a hit rate of 11% for all the objects, which is evidently high for GIS applications, query Q2 with a *combined inside- und outside-relationship* as query predicate, needed about 30ms per operation call and a time of 192 ms per hit. Query Q3 with the overlap-relationship came off better with a time of about 23ms and about 167ms per hit, respectively. Astonishing is the high percentage (80%) of the triangles that were spared with the bounding-box test. This can only be explained with particularly favourable position of the objects, i.e. a small overlapping in average. If we multiply the operation times t_{op} by 1000 (number of the objects), we maintain the time needed for the complete query. This total times are 9.51s for the distance query, 29.67s for the *combined inside- and outside-query* and 23.19s for the *overlapping query*. Especially for high numbers of hits these times let it be advisable to use a spatial access method.

It should be emphasized, however, that the hit rate for queries of typical GIS applications like point queries with "clicking of the mouse" or small area queries are clearly smaller. For a hit rate of 1%, for example, the times for Q2 and Q3 would be in a one-second interval, which should be acceptable for a GIS user.

6.4.3 Comparison with a Boundary Representation

Table 6.5 summarizes the results of measurements with the e-complex representation in comparison with a boundary representation[1] by means of the *overlap-relationship* and the data sources 2a and 3a (cuboids) as well as 2b and 3b (cylinders), respectively.

Data source	Kind of objects	Hit rate [%]	t_{op} [ms] (Brep, ECOM)	Reduction [%] (ECOM)
2a	Brep-cuboids	2	20.21	
3a	ECOM-cuboides	2	11.71	45
2a	Brep-cuboids	5	23.60	
3a	ECOM-cuboids	5	14.80	50
2a	Brep-cuboids	10	27.95	
3a	ECOM-cuboids	10	20.71	50
2b	Brep-cylinders	2	24.35	
3b	ECOM-cylinders	2	12.42	45
2b	Brep-cylinders	5	26.48	
3b	ECOM-cylinders	5	16.56	45
2b	Brep-cylinders	10	33.75	
3b	ECOM-cylinders	10	27.23	45

Table 6.5 Comparison of the Brep- with the ECOM-representation by means of the overlap-relationship with cuboids and cylinders as data source

The times for the *overlap-algorithm on the triangulated cuboids and cylinders,* respectively show that the decomposition of the objects into triangles affects positively the runtime behaviour. The number of the steps to be executed in the algorithms could be reduced essentially. Against that, in the *overlap-algorithm* every face of the first object had to be tested against every face of the second object *of the boundary representation.* This resulted in a higher number of intersections. The most intense effect was to be seen with the intersection for cuboids[2]. The hit rate was 2% and the time reduction amounted to nearly 50% for the benefit of the e-complexes. Thus we can conclude for the test with a cuboid that the objects in 3D space were grouped in that way that a minimal higher number of triangles had to be examined. The tests lead to a hit rate of 5% and 10%, respectively. As we can see from the lower part of table 6.5, the time differences are strengthened for more complex objects (cylinders with 36 triangles). For the e-complex representation as well as for the boundary representation the times for the cylinders are about 20% over those for the cuboids. Fig. 6.1 graphically shows the results from table 6.5.

---

1. A we explained above, the implemented boundary representation does not include topological information like the neighbourhood of two faces.
2. As object for the comparison query, a cuboid was used as well.

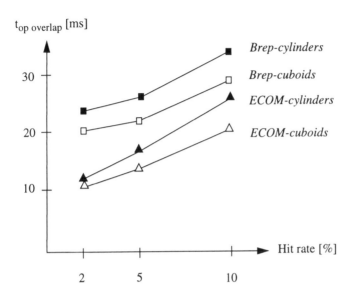

Fig. 6.1 Average time of the overlap-relationship as a function of the hit rate for cuboids and cylinders in the Brep- and the ECOM-representation

The results of the comparison tests back up the use of a spatial representation, which is based on e-complexes. The decomposition of the objects into triangles and the resulting spatial pre-selection upon few small objects (triangles) affected a clear advantage in the runtime. This is true for 2D objects in 3D space as well as for 3D objects in 3D space for the most of the measured geometric and topological algorithms. This advantage in the runtime could be increased for several algorithms with the *circle method*, i.e. the internal exploitation of the topology of the triangle networks. The number of spared triangle operations is depending on the position of the objects to each other. This is also true for the sparing with the *bounding-box test*. Altogether the e-complex in the realized implementation has proved to be a suitable representation for the examined data source in a 3D-GIS. In the next chapter we will introduce a geological application for e-complexes.

Chapter 7

A Geoscientific Application

In this chapter the practical use of e-complexes in 3D space is demonstrated by means of a geological application in the Cenozoic Lower Rhine Basin. The requirements of geology[1] to a GIS like the 3D-modeling of complex strata- and volume models show the special qualification of the e-complex for the spatial representation of geologically defined geometries. The embedding of e-complexes in an object model and database queries for geologically defined geometries of the application, demonstrate the special suitability of e-complexes for the management of geo-objects. Aiming at a 3D query support we finally present the cooperation of the query component of a prototype information system for the management of geologically defined geometries with a 3D visualization tool.

7.1 The Examination Area in the Lower Rhine Basin

The examination area includes the Lower Rhine Basin and the bordering Rhenish Massif. The Lower Rhine Basin originated on the basement of the Rhenish Massif as an intracontinental riftgraben system during a geological history of roughly 30 million years. Seen from its earth history it provides interesting requirements for geoscientific examinations. As numerous wells are drilled and the open lignite mines (like Hambach) of the RHEINBRAUN AG are situated in the middle of the examination area, the necessary data for a GIS project are available due to the courtliness of that company. Hitherto the examinations were concentrated on the cross sections in the area of the Erft block (see fig. 7.1). The examination area is a 40 x 40 km square. Eight stratigraphic horizons up to a depth of about 600 m are considered.

---

1. The listed requirements were collected in a cooperation of geologists and computer scientists within the special research project SFB 350 at the University of Bonn.

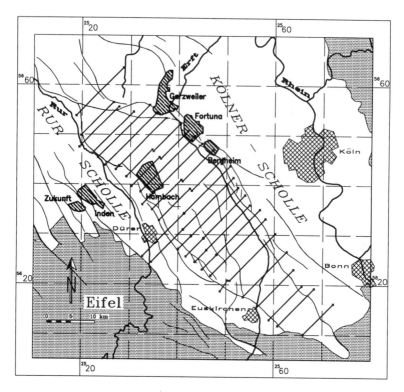

Fig. 7.1 The southern part of the Lower Rhine Basin with the important open lignite mines from the area of
the Erft block, the fault pattern and the digitalized cross sections[1] (From: ALMS, KLESPER, SIEHL
1994)

For the modeling of the data and their mapping into a database, the computer scientists and
the narrowly cooperating geoscientists are likewise challenged. A good example is the design of a geological geometric model of the Erft block, part of the Lower Rhinegraben system, as well as its adequate mapping and management in an object oriented database. The
geological data described in the following paragraphs are derived from digitalized cross sections. They are marked in the map as straight lines (see fig. 7.1). The curved lines being
roughly vertical in the cross sections are geological faults. Names of rivers and towns provide an orientation within the examination area.

---

1. For a more detailed description of the examination area see KLESPER (1994).

7.2 Requirements of Geology for a GIS

Seen from information technology, geology is characterized by dealing with three dimensional objects in their spatial and temporal development (quasi 4D). Revealing earth history can be regarded as watching a backward running film. Starting point of the examinations is the present condition of substance distributions and structures of the lithosphere. Hence the 3D-modeling as well as the spatial and temporal management of geologically defined geometries stand at the beginning of the quantitative consideration. The geometric analysis of recent geological strata and volumes is the key for the investigation of the kind and the interaction of former geological processes (SIEHL 1993). Furthermore, consistent geometric 3D models are basic requirements for the construction of geological maps (SIEHL 1988, 1993). The models can for example be intersected with digital elevation models to improve the maps or to define marginal conditions for 3D transport models of fluids. The modeling and management of time becomes necessary e.g. to execute a balanced threedimensional backword modeling (SIEHL 1993). The same is true for dynamic simulations in the sense of a forward modeling. The computational expense for the GIS is growing, if the third and the fourth dimension (time) are included. That is why the GIS more and more must be supported by external tools. Therefore the GIS occupies the role of an *integration kernel* in a federation of geoscientific tools.

7.2.1 3D-Modeling and Visualization of Geologically Defined Geometries

Before we enter into the procedure[1] for the modeling of geologically defined geometries for a GIS, we are introducing the most important geological data. In the following, sections are to be understood as interpretated data that are e.g. derived from boreholes or outcrops.

- Sections

The *sections* describe a vertical intersection through the geological structure of an examination area, i.e. the assemblage of the geological surfaces and bodies including geological *faults*. A cross section through the series of sediments consists of a set of *stratigraphic lines* that can be dislocated (faulted strata). Their geometry is given by a point set in 3D.

- Stratigraphic boundaries

The geometry of the *stratigraphic boundaries* is based on the *stratigraphic lines* in the *sections* where the stratum occurs. The stratigraphic surfaces are spread between the cross sections. E_2-complexes in 3D result from the triangulation of the *bedding-surfaces* between the point sets of the cross sections. The bedding surface being directly above or below a stratum is called *hanging stratum* and *underlying stratum* of the bedding area, respectively. The section database (point clouds) and some modeled faults are illustrated in fig. 7.2.

---

1. The general procedure is a modeling starting from primary data. The preparation of digitalized cross sections of the RHEINBRAUN AG, however, made possible a modeling being directly based on the sections.

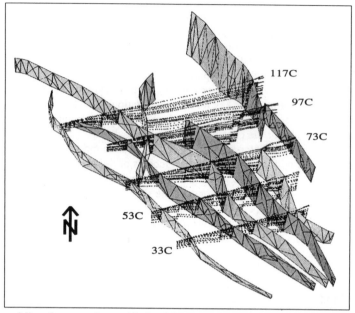

Fig. 7.2 3D-modeling of some sections and faults of the southern part of the Erft block with aid of point data
and triangulated surfaces (from: ALMS, KLESPER, SIEHL 1994)

- Faults

Faults are surfaces along which the tectonic deformation of the sediments took place.The
geometry of a fault consists of the *fault lines* of the sections through which the fault is running. Furthermore, it consists of the e_2-complexes in 3D space which are generated with the
triangulation of the fault surfaces. The modeling of the strata and faults from cross section to
cross section allows the description of interpolated geometries, lying between two cross
sections.

The procedure for the modeling of geologically defined geometries for a GIS can be described in the following three steps, taken the example of the Erft block (see also KLESPER
1994):

   1. Step: Generation of lines from digitalized point data;

   2. Step: Generation of surfaces from line data;

   3. Step: Generation of volumes from surface data.

In the first step point data of the strata and faults lying on each section i.e. the e_1-complexes,
were generated from digitalized cross sections[1] (see point clouds in fig. 7.2). This means that
the single points of the patches were composed to complete lines. In the second step surfaces

were spread between the stratum- and fault lines of different sections. The surfaces were "filled" with e_2-complexes by a triangulation[1]. The surfaces were modeled as networks with an internal topology i.e. the neighbourhoods of the triangles. In fig. 7.2 fault surfaces are illustrated that are triangulated by this procedure. Finally in the third step, the stratum- and fault surfaces are to be seen as boundaries of geologically defined volumes. They are "filled" to e_3-complexes by a tetrahedralisation[2]. Thus the local topology consists of the neighbourhoods of the tetrahedra.

In geology usually complex and irregular geometries are modeled as a mapping of their originals in nature. Thus geology aims to be very precise concerning the 3D-modeling and the *three dimensional visualization of complex surfaces* (e.g. intersection figures of arbitrarily curved surfaces) *and volumes* (SIEHL 1993). This includes the demand for an *interactive* man-machine *interface,* so that the perceptions gained during the modeling process can be integrated into the geological model.

7.2.2 Consistent Management of Geological Objects

During the modeling of geologically defined geometries a large amount of data is accumulated. Thus a handling in a database management system (DBMS) is suggested. The efficient spatial access on geological objects and efficient geometric 3D operations like the intersection of strata and faults, are examples of requirements to an object oriented query language of a GIS. Furthermore, the *browser functionality* ("which objects are in the database and which values are attached to their attributes") should be provided by the DBMS. Geometric 3D operations are the precondition for an efficient management of typical spatial queries like "provide all faults which are inside a specified box" or topological queries like "which strata build the hanging strata of stratum 7A?" The explicit representation of the internal topology of geological objects leads to further requirements to spatial access methods and geometric operations. For example a combined geometric/topological query could be: "provide the neighbours of the triangles of a fault surface from the intersection line with the stratum 7A on". For an efficient execution of this query a spatial access method must access to single triangles of the e-complex. In this example the location of the starting triangles is determined by the result of a geometric operation, i.e. the intersection of a fault with a stratum.

The *checking of the consistency of geometric volume models* is essential for the geometric 3D reconstruction, which is seen as an important first step for the reconstruction of the interaction of geological processes (SIEHL 1993). That is why the DBMS must automatically provide *spatial integrity constraints* either for the modeling of arbitrary intersections, for the safety of the quality of digitalized data or for the updating of the geometric models. In chapter 5.1 we presented classes and examples of spatial integrity constraints for the geometric modeling in geology. The extension to 4D additionally leads to the requirement of spatial queries "through time" which will be illustrated in the next section.

---

1. The cross sections were kindly provided by the RHEINBRAUN AG.
1. For optimal criteria of triangulations see KLESPER (1994).
2. This step points to future work.

7.2.3 Support of Time and Open GIS

The procedure to examine geologically defined geometries during the backward- and the forward modeling at discrete time steps leads to the demand of a *version management for paleogeological scenarios*. We can distinguish two ways to manage geological objects at discrete time steps in a database. First a management of different versions of a single geologically defined geometry object at different discrete time steps has to be provided for the support of the geological backward modeling. Second, the *management of configurations,* i.e. the handling of different geometries of one geological object during the same time, plays an important role in the *interactive design process.* With his background knowledge the geologist finally can interactively select the most plausible configuration. Both requirements lead to the necessity of the *support of* so called *long database transactions* during the geological design process, i.e. the data has to be checked out for the processing in main memory and later the best version has to be checked in the database again. One of the possible implementations is the use of temporary versions for objects. As a vision, today, geologists already require the embedding of geodynamic processes as "real" 4D components into the data handling of a GIS. This, however, also presupposes the real integration of the data modelling and the process modelling, which still is an unsolved problem. In future, object based techniques like the object modeling technique by RUMBAUGH (1991) could help to build a bridge across the two fields.

The inclusion of time leads to an increasing expenditure of computing in the GIS. Therefore existing *autonomous operation services* (SCHEK and WOLF 1992) should be used for costly computations. Thus a reimplementation of the functionality needed in the GIS is avoided. The GIS must be open concerning the communication to foreign systems. It must include exact knowledge of the provided services in the system federation and of their interfaces like a data broker. An important requirement in this context is the *integration of heterogeneous 3D data sources,* to which this book has tried to make a contribution.

7.3 Database Support for a GIS in Geology

7.3.1 Object Oriented Modeling of Geologically Defined Geometries

An object oriented modeling allows the mapping of hierarchies of geological objects as they typically occur in the *stratigrBoundary-subStratumBoundary* relationship or between the geometric objects *FaultSurface-Triangle-Point* (see fig. 7.3). We introduce an object model[1] of geologically defined geometries as it is realized in GEOSTORE (BODE et al. 1994). The most important object classes, namely cross section, or for shortness *section, stratigrBoundary and fault* we already introduced in chapter 7.2.1.

---

1. Essential aspects of the object model were designed in cooperation with the Geological Institute of the University of Bonn (research group of Agemar Siehl) within a database practical course under the guidance of Thomas Bode.

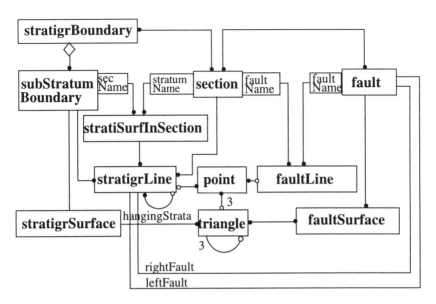

Fig. 7.3 OMT object diagram of geological objects

Fig. 7.3 shows the *OMT object model* (RUMBAUGH et al. 1991) of the geological objects realized in GEOSTORE[1]. Every *subStratumBoundary* is part of a *stratigrBoundary* and has a set of *stratigrLines* and a *stratigrSurface* as geometry[2]. A *stratigrLine* has a reference to the section on which it is lying, to the *subStratumBoundary* to which it belongs to and to the *leftFault* and the *rightFault* which are bordering the *stratigrLine*. A *stratigrLine* is vertically bounded by its *leftFault* and its *rightFault*. From the class *stratgrSurfInSection* one may access efficiently on all *subStratumBoundaries* of a *section* and all *sections* of a *subStratum-Boundary* over the *section-* and the *stratumName,* respectively. A *stratigrLine* consists of a list of *points*. Furthermore, the *hangingStrata* of all *stratigrLines* consists of a set of *stratigrLines*. An instance of a *fault* represents a *Fault* in all sections of the examination area with references to their *faultLines* and *faultSurfaces*. *stratigrSurfaces* and *faultSurfaces* are modeled as a list of *triangles,* respectively. They consist of three points and their neighbouring triangles. Every FaultLine has a pointer over the *faultName* to the *section* and the *faults,* respectively, to which it or they belong. Hitherto in the object model (fig. 7.3), only the first two steps of the modeling of geologically defined geometries are reflected, i.e. the generation of lines from point data and the generation of surfaces from line data. Fig. 7.4 shows a possible extension of the object model. Stratigraphic and fault surfaces are represented as triangle networks which are "filled" up to convex tetrahedron networks (convex e_2-complexes)[3].

---

1. The implementation of the base functionality of GEOSTORE was supported by the students of the mentioned database practical course. As an implementation base we used the ODBMS "ONTOS" (1992). ONTOS DB is a registered trademark of ONTOS Inc.
2. In future, additionionally a *StratigrVolume* will be provided. We use the term "stratum" as a general term like section and fault, independently of its geometry. I.e. an object of the class *stratum* may contain lines, a surface or a volume as geometry.

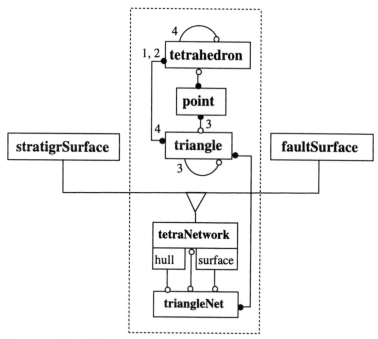

Fig. 7.4 Extension of the OMT object model for geological objects

Besides the new class *tetrahedron,* which has instances with references to four *triangles,* we have added the classes *triangleNetwork* and *tetraNetwork.* In the extended model *stratgrSurface* and *faultSurfaces* consist of two *triangleNetworks,* the convex *hull* and the "real" *surface.* Additionally, they have a *tetraNetwork* which fills the space between the *hull* and the *surface* with tetrahedra.

7.3.2 Management of the 3D Geometry and Topology of Geological Objects

We divide database queries on sections, strata and faults into the two categories *topological* and *geometric queries.* For a better illustration we give some examples that for the most part are implemented in GEOSTORE.

1) *Topological queries*

With reference to the terminology in chapter 4, we mean by "topological queries" either database queries that relate to the *local topology* of single geologically defined geometries or such queries that aim at the *global topology.* The result of the latter queries is the relative topological situation between different geological defined geometries.

---

3. Fig. 7.4 shows an *internal* extension of the object model which gives a hint for an efficient *implementation of stratum and fault faces.* A further step would be an *external* extension of the object model by representing *strata* as volumes, i.e. tetrahedron networks (convex e_3-complexes).

Queries to the local topology:

Examples for this type of queries are:

- "Provide all triangles with their neighbourhoods for the top stratigraphic surface of the stratum 6A";
- "provide all triangles with their neighbourhoods for the fault surface of the fault ROEV0";
- "provide all triangles with their neighbourhoods for the stratigraphic surfaces that are lying between the faults ROEV0 and WITT0";
- "provide all triangles with their neighbourhoods of the fault ROEV0 within a specified box";
- "provide all triangles with their neighbourhoods of the fault ROEV0 that are lying on the positive and the negative side, respectively, of the intersection line with stratum 6A" (see fig. 4.15).

As the local topology, i.e. the neighbourhoods of the triangles, are explicitly given in the object model, it can be easily accessed. Local topological queries may be combined with geometric queries, as we have seen in the last example above. In this example the result of the query is the local topology, starting from an intersection area.

Queries to the global topology:

Examples for queries to the global topology are:

- "Does the fault ROEV0 intersect the stratigraphic surfaces 6A, 6B and 6C?";
- "provide all the strata that run through the section 117C";
- "provide the hanging strata and the underlying strata of the stratum 6A";
- "provide the upper surface of the stratum 6A between the faults ROEV0 and WITT0".

Some topological relationships, like *"all the faults running through a section"*, are explicitly provided in the introduced object model (see fig. 7.3). Other relationships, like *the intersection between faults and strata,* however, must be dynamically computed during the query. It always is a matter of discretion which topological relationships are or are not modeled "hard wired" in the object model. Ultimately the user has to decide individually from case to case according to the frequency of the respective queries.

If updates also concern the boundary of the topology, the updates of the topology of single geological objects may influence neighbouring topologies as well. That is why one should demand as an *integrity constraint* that the boundary of the topology to be updated must not be changed.

2) *Geometric queries*

It is characteristic for geometric database queries that new geometries are computed from the existing geometries in the database. An example is the intersection of a stratigraphic surface with a fault surface. In this case an intersection line is computed as a new geometry.

Examples for geometric queries are:

- "Provide the intersecting geometries of the specified horizontal plane with the triangles of the stratum and fault surfaces" (see fig. 7.5);
- "provide the intersecting geometries of the specified vertical plane with the triangles of the stratum and fault surfaces" (see fig. 7.6);
- "provide the intersecting geometries of the intersection between the fault ROEV0 and the stratum 6A".

Fig. 7.5 Horizontal intersection through the fault pattern of the Erft block in a depth of 100m

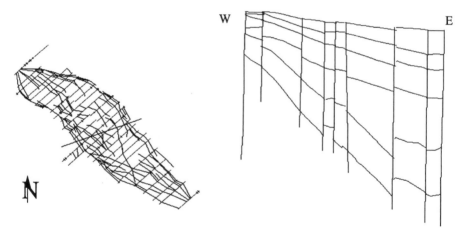

Fig. 7.6 Vertical intersection through the geometric elements of the Erft block of the Lower Rhine Basin that are represented in the database. Overview and section map.

If we allow flexible intersection operations (see fig. 7.5 and 7.6) in that way that the intersection plane may be arbitrarily chosen, these geometric operations cannot be considered "hard wired" in the object model, but must be dynamically computed during the running time of the query evaluation. The GIS requirement of three dimensional spatial queries directly leads to the question how the user can be efficiently supported during the query formulation. In the next chapter we follow the answer of this question.

7.3.3 3D-Visualization and Coupling with GIS Tools

Although today's database management systems provide a comfortable support of the query formulation with object browsers and though SQL has developed to a quasi standard query language for databases during the last decade, it seems to be unalterable - especially for spatial queries - to additionally provide a *visual 3D support for the query formulation.* An example from our geological application is the following: consider, you intend to select all strata that are running through a certain section and that are lying between a given left and right fault. Intuitively, you would like to formulate the query with a graphical selection of the section and the faults within a 3D volume model. With the adding of the time parameter (4D) one could even visualize the development of geological structures in time. Thus, one could pose queries in *virtual reality* by "stopping the time" during the running of the query *(walking through space and time).* Of course this is only possible f we had stored a series of 3D-models, i.e. discrete time intervals, in the database. A first small step in this direction is the version management of geologically defined geometries mentioned in chapter 7.2.3.

If we generally consider the architecture of the GIS, the question arises, if the visualization of geo-objects should be executed in the GIS itself or in an external tool. One of the following three realizations may be chosen:

(1) narrow coupling of a visualization tool with a GIS;
(2) loose coupling of a visualization tool with a GIS;
(3) open GIS environment.

Realization (1) has the advantage that the user only has to know one single system and that no high costs arise for the data transfer between the data management and the visualization within the system. Furthermore, the object management of the GIS may set access to primitive graphics objects like points or lines. Thus a *unified object model* is used for the data management and for the visualization. Additionally, the question of the conversion between different data formats of the GIS and of the visualization tool is superfluous within a narrow coupling. The disadvantage, however, is that the narrow coupling between the data management and the visualization leads to a dependency that may negatively affect updates of the data structures in the GIS or in the visualization tool.

A loose coupling (realization 2), however, is more flexible for the developer and allows the coupling of several GISs with one visualization tool over a communication protocol, like remote procedure call (RPC) or CORBA[1] (1991) (Common Object Request Broker Architecture). However, one should take care that the preselection of the objects to be visualized (window query) isexecuted in the GIS and not by the graphics functions in the GIS. Thus the goal should be to reduce the data transfer across the network.

---

1. Common Object Request Broker Architecture. CORBA is a registered trademark of the OMG (Digital Equipment Corporation, Hewlett-Packard Company, HyperDesk Corporation, NCR Corporation, Object Design Inc., SunSoft Inc.).

Realization (3), the open GIS environment, enables the developer to write GIS- and visualization functions that may be narrowly or loosely coupled. The user may use a toolkit that is extensible for new "building blocks". Thus the disadvantage of being unflexible (realization 1) is reduced by a *modular architecture of the GIS*.

In the following, we introduce a combined solution of realization (1) and (2) as it is realized in GEOSTORE. Hitherto, GEOSTORE provides a twodimensional visualization in the system itself[1] and a threedimensional, externally realized visualization. The 2D and 3D visualizations could also serve as building blocks of an open GIS environment. For the 3D visualization, the 3D visualization and modeling tool GRAPE[2] (SIEHL et al. 1992; GRAPE 1993) was coupled with GEOSTORE. Fig. 7.7 shows a part of a triangle network of stratum and fault surfaces visualized with GRAPE.

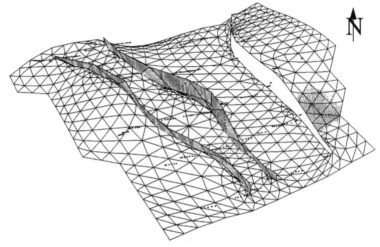

Fig. 7.7 3D visualization of stratum and fault surfaces of the Erft block as triangle network and cross sections as point clouds (from: KLESPER 1994)

For the realization of the coupling we used a program library[3] that was developed at the Institute of Computer Science III of the University of Bonn. It provides a data exchange over network on different spatially distributed hardware platforms and an access to external services. At the application layer, the program library provides the usual procedures for the initialization, opening, closing of the communication channel and for the sending and receiving of messages. Fig. 7.8 shows the starting of a typical client server connection between the visualization tool (GRAPE) and the object oriented geo-database system (GEOSTORE) on two different hardware platforms.

---

1. Realized with UIT (an object oriented shell) on top of CGI, a graphics interface for UNIX with X-Window System support. The X Window System is a registered trademark of the Massachusetts Institute of Technology. UNIX was a registered trademark of AT&T, now of X/Open.

2. GRAPE was developed within the special research project at the Dept. of Applied Mathmematics of the University of Bonn and is used at the Geological Institute for the computer supported construction of threedimensional geological strata and volume models.

3. The implementation of the protocol is realized between the transport and the application layer of the ISO/OSI layer model and is based upon UNIX ports and sockets.

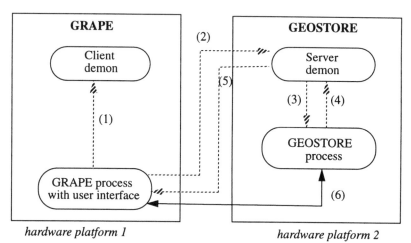

hardware platform 1   *hardware platform 2*

Fig. 7.8 Starting a connection between GRAPE and GEOSTORE

First, the GRAPE process (with user interface) has to registrate itself at the client demon (1).
To send a message at the GEOSTORE process, a query for the communication address of the
GEOSTORE process has to be send (2) and the GEOSTORE process has to be started from
the server demon (3). After that the started GEOSTORE process registrates itself at his local
demon and tells it its communication address (4). Now the server demon may send the communication address of the GEOSTORE process to the GRAPE process with user interface
(5). Thereupon the connection between the server demon and the GEOSTORE process can
be terminated. Finally, the GRAPE process can enter a connection with the GEOSTORE process (6) and exchanges data with the GEOSTORE process.

The coupling between GRAPE and GEOSTORE points out a way how GISs may be coupled
with other tools used in the geosciences. Aiming at a cost-effective data exchange from and
to the GIS, in future compression techniques as they where addressed for topological structures in chapter 4.4.2, will gain importance. The same will become true for the use of techniques from the field of interoperable information systems (CREMERS et al. 1992; KNIE-
SEL et al. 1991). Hereby the DBMS of the GIS may serve as a *GIS-database server* (EB-
BINGHAUS et al. 1994) that handles spatial information e.g. for *interoperable GIS* (SCHEK
and WOLF 1993). Because of the good approximation of the geometry, the GEO-complexes
are a suitable representation for the visualization that allow a flexible interactive intervention
into the geological modeling in the points of the triangle- or tetrahedron-networks.

Chapter 8

Summary and Outlook

An essential theme of our examinations was to develop a model for the integration of spatial information for 3D Geo-Information Systems. As a central problem the integration of different spatial representations had to be solved. We first analysed 2D representations known from GISs and the 3D representations of CAD systems. It showed up that none of these representations is suitable for an integrated representation of the geometry and the topology of geo-objects in 3D-GISs. With approximations and topological abstractions, ways for the integration of spatial information were pointed out. They lead to a three-level notion of space that likewise considers geometry, metrics and topology. We could attach types of spatial operations to each of the three layers that were embedded as "basic building blocks" in spatial queries of a GIS.

Unified spatial representation

We have introduced the e-complexes as a unified representation for the management of geometry and topology in 3D-GISs. They are supporting the different layers of the notion of space. The approach of a unified representation has the advantage that a user does not have to convert all representations to each other, for instance to process them afterwards. Only n transformations have to be executed, if n different representations exist. The transformation from the 3D-representations into the e-complex representation results from a triangulation of the surface of the 3D-objects and a subsequently tetrahedralisation of their volumes. Furthermore, in the e-complex the topology is explicitly represented. Thus topological operations can be executed that cannot be provided or that are very costly in the special representations like B-Rep or octree etc. That is why the "general system", i.e. the integrated GIS, obtains an extended functionality in comparison to the "specialized systems".

Foundations of the ECOM-algebra

We started from the simplicial complex that consists of a topological structure and is composed of simplices, i.e. nodes in 0D, edges in 1D, triangles in 2D and tetrahedra in 3D. The spatial objects of the ECOM-algebra are e-complexes, i.e. simplicial complexes that are extended by a local topology (that may contain holes) and by a geometry. We have extended the algebra for binary topological relationships for 2-simplices with codimension 0 by Egenhofer in 3D space with arbitrary codimension. For the specification of topological operations additionally the local topology of the objects was considered. Thus, refined configurations of topological relationships could be distinguished. In the global case the topological relationships could be specified by the intersection of the interior and the boundary of those lower dimensional interior simplices that participated in the intersection. In the local case, the topological relationships could be directly specified on the internal topology of the objects. Besides topological operations, the ECOM-algebra contains as well metrical operations and direction operations that could be extended for not-connected e-complexes.

For an integration of thematic and spatial attributes we have extended the e-complex with thematic attributes and called it "GEO-complex". Two methods for the treatment of thematic attributes during the intersection of e-complexes were demonstrated by the example of geologically defined geometries. For the management of GEO-complexes we have introduced building blocks of a GEO-model kernel for open GISs. The GEO-model kernel has a modular and open system architecture and provides two kinds of extensible building blocks, namely CASE and service building blocks. The embedding of the building blocks into extensible database kernel systems was explained with the examples of the DASDBS Geokernel and the Object Management System (OMS).

Efficient geometric algorithms

The introduced geometric algorithms are characterized by the "breaking down" of spatial operations upon single simplices. Thus we could use a more efficient bounding-box test on a reduced set of objects. Furthermore, the explicit exploiting of the topology of the e-complexes improves the running time of the algorithms. Especially the last point could be used for inclusion and intersection operations.

We have introduced the *overlapOp*-algorithm for two convex e-complexes with running time[1] $O(s)$, let s be the number of intersections of the d-simplices of both e-complexes. In this algorithm a circle-like surfing over the single simplices, i.e. the explicitly stored neighbourhood relations (local topology) of the e-complexes, is provided. With the help of the convex geometry of the e-complexes, the start of the algorithm could be reduced to the search of a single starting simplex. An efficient access upon the starting simplex may be guaranteed by a qualified spatial access method. We have introduced algorithms for the computation of the convex hull, i.e. the generation of convex e-complexes for e_2-complexes in 2D space in $O(m)$ time and for e_2- and e_3-complexes in 3D space in $O(nlogr)$ time, m being the number of triangles (2-simplices) generated during the algorithm and r being the number of points (0-simplices) of the convex hull.

Performance tests

Performance tests of geometric and topological operations on e-complexes were executed within the extensible database kernel system OMS. Geologically defined 3D-areas served as data as well as artificially generated boxes and approximated cylinders, generated by rotation and clipping. The running times for the operations on e-complexes for the used triangle networks amounted to about 100 ms with exception of the overlap operator, which generates an intersecting object and which in the realized version did not yet exploit the internal topology. It was shown up that in most cases the decomposition of the objects into triangles caused a

---

1. The time is valid for the actual algorithm after the search of the starting triangles. In the worst case the search of the starting triangles is $O(n^2)$. Let n be the number of triangles and tetrahedra of the e-complexes, respectively.

clear improvement of the running time for the bounding box test, which was executed as a spatial preselection. In this way 50% of the needed triangle comparisons could be spared in average. The sparing, however, is dependent on the degree of the overlapping of the data. The _inside-algorithm_ for triangulated areas which was realized by the _circle method,_ came off 25% better than comparable algorithms without exploitation of the topology. In comparison with a boundary representation, the overlap algorithm on the e-complex representation came off 7-50% better, depending on the hit rate of the qualified objects.

Geoscientific application in the Lower Rhine Basin

A geological application from the Lower Rhine Basin showed that the introduced geometric and topological representation, which is based on e-complexes may be used in practice for the modeling and management of geological strategraphic surfaces and volumes. Boundary surfaces of geological strata and faults were modeled as e_2-complexes. Examples of geometric and topological queries were introduced by means of GEOSTORE, an information system for the management of geologically defined geometries. The visual query support of spatial queries and their results turned out to be an important requirement. This was practically demonstrated with the coupling between GEOSTORE and GRAPE, a 3D visualization tool for the construction and modeling of geological strata models.

Outlook

We introduced a model for the integration of spatial information and evaluated essential parts of it for a geological application. A next step could be to realize the whole model., inclusively e_3-complexes, as a building block of the GEO-model kernel and to evaluate it by means of a geological application in GEOSTORE. Hereby the following future questions arise:

- The extension of the area modeling (e_2-complexes) to a _complete volume modeling (e_3-complexes)_ in the geological application. Unclear is especially the connection of thematic and spatial attributes on GEO-complexes under consideration of the results of geometric and topological operations. Hitherto only the integration of thematic attributes with the nodes of e-complexes was discussed from an application point of view. New requirements, however, could result, if attributes were also introduced at the edges or at arbitrarily parts of surfaces through interpolation. A special problem then is the intersection of such GEO-complexes.

- A further interesting question is the _optimization of spatial queries._ For that purpose the user should have a cost model at his disposal, which should estimate the costs for the execution of geometric operations and the spatial access on e-complexes.

- An extension of the model to _4D_ is a new challenge. The management of different geometry versions for the modeling of geological processes is a first step in this direction. Also unsolved till now is the question how the update of the topology between two scenes (snapshots) can be exploited for a realtime visualization of geoscientific processes.

With our model and its application a gap should be closed between hitherto existing pure topological approaches like those of EGENHOFER or PIGOT and the modeling of geometry and topology in today's GISs that are restricted to 2D space. It remains to be seen how fast the threedimensional modeling as well as complex geometric and topological operations on 3D objects will meet with commercially available GISs. The tendency already shows into this direction with the development of more flexible and first object oriented GISs in opposite to the hitherto existing layer based GISs. Certainly further optimizations like parallel geometric algorithms will also gain in significance. This could lead to an increase of the acceptance for GISs among the users. The GIS technology could even take over a leading role for other spatial applications like environmental sciences, CAD, medicine or robotics. This could have a feedback effect e.g. for the development of 3D-CAD systems. This hypothesis is strengthened by the fact that existing 2D-GISs can already execute some geometric operations (e.g. map overlays) with a connection of thematic and geometric data which is not the case at today's CAD systems. The significance of a *unified spatial representation* for the data handling in GIS will certainly increase because of a strengthened *use of databases* (DBMS/GIS coupling) and because of the growing *information highways* and the herewith connected *interoperable use of GIS*.

References

ABEL D (1988) Relational data management facilities for spatial information systems. In: Proc. 3rd Int. Symposium on Spatial Data Handling, Sydney, Australia, pp 9-18

ABEL D, Ooi BC (eds, 1993) Advances in Spatial Databases, Proceedings of the Third International Symposium, SSD'93, LNCS, Springer, Berlin Heidelberg New York

ABEL DJ, WILSON MA (1990) A Systems Approach to Integration of Raster and Vector Data and Operations. In: Proc. 4th Int. Symposium on Spatial Data Handling, Zürich, Vol. 2, pp 559-566

ABEL DJ, YAP S K, WALKER G, CAMERON MA, ACKLAND RG (1992) Support in Spatial Information Systems for Unstructured Problem-Solving. In: Proceedings of the 5th International Symposium on Spatial Data Handling (SDHS), Charleston, South Carolina, pp 434-443

ACM (1990) Special Issue on Heterogeneous Databases, ACM Computing Surveys 22 (3)

AdV (1989) ATKIS-Gesamtdokumentation, Arbeitsgemeinschaft der Vermessungsverwaltungen der Länder der Bundesrepublik Deutschland (AdV)

ALMS R, KLESPERS C, SIEHL A (1994) Geometrische Modellierung und Datenbankentwicklung für dreidimensionale Objekte. In: IfAG Nachrichten aus dem Karten- und Vermessungswesen, Series 1

ANDERL R, SCHILLI B (1988) STEP - Eine Schnittstelle zum Austausch integrierter Modelle. In: H R Weber (Hrsg). CAD-Datenaustausch und -Datenverwaltung, Springer-Verlag Berlin, Heidelberg

BAK PRG, MILL AJB (1989) Three dimensional representation in a Geoscientific Resource Management System for the minerals industry. In: J Raper (ed), Three Dimensional Applications in Geographic Information Systems, Taylor & Francis, London, pp 155-182

BANCILHON F, CLUET S, DELOBEL C (1989) A Query Language for the O_2 Object-Oriented Database System. In: Proc. of the 2nd Workshop on Database Programming Languages, Salishan, Oregon

BARBER CB, DOBKIN DP, HUHDANPAA H (1993) The Quickhull Algorithm for Convex Hull, Geometry Center, University of Minnesota, Minneapolis, Technical Report GCG53

BARTELME N (1989)   Norbert Bartelme, gis Technologie, Springer, Berlin Heidelberg New York

BATORY D, BARNATT J, GARZA J, SMITH K, TSUKUDA K, TWICHELL C, WISE T (1986) GENESIS: A Reconfigurable Database Management System, Tech. Rep. 86-07, Dept. of Comp. Science, Univ. of Texas at Austin

BAYER R, McCREIGHT E (1972) Organization and Maintenance of Large Ordered Indexes. In: Acta Informatica 1, Springer, Berlin Heidelberg New York, pp 173-189

BECKMANN N KRIEGEL H-P, SCHNEIDER R, SEEGER B (1990) The R*-tree: An Efficient and Robust Access Method for Points and Rectangles. In: Proceedings ACM SIGMOD, Atlantic City, N.Y., pp 322-331

BERKEL T, KLAHOLD P, SCHLAGETER G, WILKES W (1988) Modelling CAD-Objects by Abstraction. Proceedings of the third International Conference on Data and Knowledge Bases, Jerusalem, pp 227-239

BILL R, FRITSCH D (1991) Grundlagen der Geo-Informationssysteme, Wichmann Verlag (2 Vol.), Karlsruhe

BODE TH, BREUNIG M, BÜTTNER G, CREMERS AB, REDDIG W (1993) Design and Construction Support with Object-Oriented Database Kernel Systems. In: Proceedings of the 5th International Conference on Computing in Civil and Building Engineering, Anaheim, California, pp 931-938

BODE, TH, BREUNIG M, CREMERS AB (1994) First Experiences with GEOSTORE, an Information System for Geologically Defined Geometries. In: J Nievergelt, Th Roos, H-J Schek, P Widmayer (eds), IGIS'94: Geographic Information Systems, Proceedings of the International Workshop on Advanced Research in Geographic Information Systems, Monte Verita, Ascona, Schweiz, Feb. 28 - March 4, LNCS 884, Springer, Berlin Heidelberg New York, pp 35-44

BODE TH, CREMERS AB, FREITAG J (1992) OMS - ein erweiterbares Objektmanagement-System. In: Objektbanken für Experten, GI Fachberichte, Springer, pp 29-54

BOUILLE F (1976) Graph Theory and Digitization of Geological Maps, Computers & Geosciences, Vol.2, pp 375-393

BOURSIER P, MAINGUENAUD M (1992) Spatial Query Languages: Extended SQL vs. Visual Languages vs. Hypermaps. In: BRESNAHAN et al. (eds), pp 249-259.

BRASSEL K (1993) Grundkonzepte und technische Aspekte von Geographischen Informationssystemen, IJK, pp 31-52

BRESNAHAN EP, CORWIN E, COWEN D (eds)(1992) Proceedings of the 5th International Symposium on Spatial Data Handling, Charleston SC, IGU Commission of GIS

BREUNIG M (1991) Räumliche Abstraktion für Planungsaufgaben im Bergbau. In: W Skala (Hrsg), Forschungsberichte zur Mathematischen Geologie und Geoinformatik, FU Berlin

BREUNIG M, BODE TH, CREMERS A B (1994) Implementation of Elementary Geometric Database Operations for a 3D-GIS. In: Th Waugh and R Healey, Proceedings of the 6th International Symposium on Spatial Data Handling, 5th - 9th September 1994, Edinburgh, Scotland, UK, pp 604-617

BREUNIG M, DRÖGE G, WATERFELD W (1990) Der DASDBS Geokern. Dokumentation des DASDBS-Geokerns, DVSI, Fachbereich Informatik, TH Darmstadt

BREUNIG M, HEYER G, PERKHOFF A, SEEWALD M (1991) An Expert System to Support Mine Planning Operations. In: G Karagiannis (ed), Proceedings of the 2nd International Conference on Database and Expert Systems Applications (DEXA), Berlin, Springer, pp 293-298

BREUNIG M, PERKHOFF A (1991) Data and System Integration for Spatial Reasoning in XPS-Applications, internal report No. 3/91, Institut für Informatik, FU Berlin

BREUNIG M, PERKHOFF A (1992) Data and System Integration for Geoscientific Data. In: BRESNAHAN et al. (eds), pp 272-281

BRINKHOFF TH, HORN H, KRIEGEL H-P, SCHNEIDER R (1993) Eine Speicher- und Zugriffarchitektur zur effizienten Anfragebearbeitung in Geo-Datenbanksystemen, BTW'93

BRUZZONE E, DE FLORIANI L, PELLEGRINELLI M (1993) A Hierarchical Spatial Index for Cell Complexes. In: SSD 93 , Signapore, LNCS 692, Springer, Berlin Heidelberg New York, pp 105-122

BURNS K L (1975) Analysis of Geological Events. In: Mathematical Geology, Vol. 7, No. 4, 1975

BURROUGH PA (1990) Methods of spatial analysis in GIS, Preface of the special issue on methods of spatial analysis in GIS. In: International Journal of Geographical Information Systems, Vol. 4, No. 3, July-Sept., Taylor & Francis

BÜSCHER K, KIRCHHOFF CH, STREIT U, WIESMANN K (1992) Vergleich der Nutzbarkeit und Auswahl von GIS für die Regionalisierung in der Hydrologie. In: U Streit (Hrsg), Werkstattberichte Umweltinformatik - Agrarinformatik - Geoinformatik, Heft 1, Okt. 1992, Inst. f. Geographie und Institut für Agrarinformatik, Universität Münster

CALKINS HW, TOMLINSON RF (1977)   Geographic Information Systems: Methods and Equipment for Land Use Planning. IGU Commission on Geographical Data Sensing and Processing and U.S. Geological Survey, Ottawa

CAREY MJ, DEWITT DJ, FRANK D, GRAEFE G, MURALIKRISHNA M, RICHARD-SON JE, SHEKTIA EJ (1988)   The Architecture of the EXODUS Extensible DBMS. In: M Stonebraker (ed), Readings in Database Systems, Morgan Kaufman

CHOU H, DING Y (1992) Methodology of integrating Spatial Analysis/Modelling and GIS. In: BRESNAHAN et al. (eds), pp 514-523

CLARKE KC (1986)   Advances in Geographic Information Systems. -Computers, Environment and Urban Systems. 10, pp 175-184

CORBA (1991)   The Common Object Request Broker: Architecture and Specification. OMG Document 91.12.1, Digital Equipment Corporation, Hewlett-Packard Company, HyperDesk Corporation, NCR Corporation, Object Design Inc., SunSoft Inc., Dec. 1991

COWEN DJ (1988)   GIS versus CAD versus DBMS What Are the Differences. In: Photogramm. Eng. Remote Sens., 54, 11, Falls Church, VA, pp 1551-1555

CREMERS AB, KNIESEL G, LEMKE T, PLÜMER L (1992)   Intelligent Databases and Interoperability. In F Belli, F J Radermacher (eds), Industrial and Engineering Applications of Artificial Intelligence and Expert Systems, LNAI No. 604, Springer

DANGERMOND J (1983)   A Classification of Software Components Commonly Used in Geographical Information Systems. In: D Marble, H Calkins, D Peuquet (eds), Basic Readings in Geographic Information Systems, SPAD Systems, Amherst, N.Y.

DANN R, SCHULTE-ONTROPP R (1989)   DIGMAP- Das Software-System im Markscheidewesen der Ruhrkohle AG

DAVID B, RAYAL L, SCHORTER G, MANSART V (1993)   GeO$_2$: Why Objects in a Geographical DBMS? In: The 3rd International Symposium on Large Spatial Databases, Singapore, 23rd - 25th June, pp 264-276

DAYAL U, MANOLA F, BUCHMANN A, CHAKAVARTHY D, GOLDHIRSCH D, HEILER S, ORENSTEIN J, ROSENTHAL A (1987)   Simplifying Complex Objects: The PROBE Approach to Modelling and Querying them. In: Proceedings BTW'87

DE FLORIANI L, GATTORNA G, MARZANO P, PUPPO E (1994)   Spatial Queries on a Hierarchical Terrain Model. In: Th Waugh and R Healey, Proceedings of the 6th International Symposium on Spatial Data Handling, Edinburgh, pp 819-834

DE HOOP S, VAN DER MEIJ L, VAN HEKKEN M, VIJLBRIEF T (1994) Integrated 3D Modeling within a GIS. In: Proceedings of the International GIS Workshop "Advanced Geographic Data Modelling", AGDM'94, Delft

DIKAU R (1992) Aspects of Constructing a Digital Geomorphological Base Map. In: From Geoscientific Map Series to Geo-InformationSystems, Geologisches Jahrbuch A122, Hannover, pp 357-370

DTV (1991) dtv-Atlas zur Mathematik, Band 1, 9. Auflage, Nov., Deutscher Taschenbuch Verlag

DYBALLA A, TOBEN H, LINEMANN V, SAAKE G (1991) Integration geometrischer Daten in ein erweiterbares Datenbanksystem, IBM-Report TR 75, 91, 12, Mai

EBBINGHAUS J, HESS G, LAMBACHER J, RIEKERT W-F, TROTZKI T, WIEST G (1994) GODOT: Ein objektorientiertes Geoinformationssystem. In: L M Hilty, A Jaeschke, B Page, und A Schwabl (eds): Proceedings 8. Symposium für den Umweltschutz, Hamburg, Metropolis-Verlag, Marburg, pp 351 - 360

EGENHOFER MJ (1989) Concepts of spatial objects in GIS user interfaces and query languages. GIS/LIS '89, Orlando, FA, USA, November

EGENHOFER MJ (1989a) A Formal Definition of Binary Topological Relationships. In: W Litwin, H-J Schek (Hrsg), Foundations of Data Organisation and Algorithms, Proceedings FODO 1989, Paris LNCS 367, Springer, Berlin Heidelberg New York, pp 457-472

EGENHOFER MJ (1989b) Spatial Query Languages. PhD-thesis, April, Univ. Maine, Orono, US

EGENHOFER MJ (1991) Reasoning About Binary Topological Relations. In: O Guenther and H-J Schek (eds), Advances in Spatial Databases, 2nd Symposium SSD'91, Zürich, LNCS 525, Springer, Berlin Heidelberg New York, pp 143-160

EGENHOFER MJ et al. (1989) A Topological Data Model for Spatial Databases. In: A. Buchmann and O. Günther (eds), Proceedings SSD'89, LNCS 409, Springer, Berlin Heidelberg New York, pp 271-285

EGENHOFER MJ, FRANK AU (1988) Designing Object-Oriented Query Languages for GIS: Human Interface Aspects, Proceedings of Spatial Data Handling, Sidney

ERWIG M, GÜTING RH (1991) Explicit Graphs in a Functional Model for Spatial Databases. Informtatik Berichte Nr. 110, 6/91, FernUniversität Hagen

FERRUCCI V, VANECEK G (1991) A Spatial Index for Convex Simplicial Complexes in d Dimensions. In: O Guenther and H-J Schek, Advances in Spatial Databases, 2nd Symposium SSD'91, Zürich, LNCS 525, Springer, Berlin Heidelberg New York, pp 361-380

FGDC9 (1994) Content Standards for Digital Geospatial Metadata, Federal Geographic Data Committee, June 8

FINDEISEN D (1990)   Datenstruktur und Abfragesprachen für raumbezogene Informatio-
nen, Schriftenreihe des Instituts für Kartographie und Topographie der Rheinischen
Friedrich-Wilhelms-Universität Bonn, Kirschbaum Verlag, Bonn

FONG EN, GOLDFINE AH (1989)   Special Report: Information Management Directions:
The Integration Challenge, SIGMOD RECORD Vol. 18, No. 4, Dez, pp 41-43

FRANK AU (1994)   Qualitative Temporal Reasoning in GIS-Ordered Time Scales. In: Th
Waugh and R Healey, Proceedings of the 6th International Symposium on Spatial Data
Handling, Edinburgh, pp 410-430

GAEDE V, RIEKERT W-F (1994)   Spatial Access Methods and Query Processing in the
Object-Oriented GIS GODOT. In: Proceedings the International GIS Workshop on Advanced Geographic Data Modelling (AGDM-94), Delft, 12. - 14. Sept., im Druck

GAVRILA DM (1994)   R-Tree Index Optimization. In: Th Waugh and R Healey, Proceedings of the 6th International Symposium on Spatial Data Handling, Edinburgh, pp 771-
791

GOODCHILD MF (1985)   Geographic Information Systems in Undergraduate Geography:
A Contemporary Dilemma, The Operational Geographer, Vol. 8, pp 34-38

GOODCHILD MF (1990)   Spatial Information Science. Keynote Adress, 4th. Int. Sympo-
um on Spatial Data Handling. Proceedings, Vol. 1, Zürich, pp 3-12

GÖPFERT W(1991)   Raumbezogene Informationssysteme, 2. Edition, Wichmann Verlag,
Karlsruhe

GRAPE (1993)   Universität Bonn (Hrsg). A Graphical Programming Environment for Mathematical Problems; Version 4.0; Institut für angewandte Mathematik, SFB 256

GRASS (1993)   GRASS 4.1 Reference Manual, U.S. Army Corps of Engineers, Construction Engineering Research Laboratories, Champaign, Illinois.

GRUGELKE G (1986)   Benutzerhandbuch THEMAK2, Version 2.0, FU Berlin

GRÜNBAUM (1967)   Convex Polytopes. Wiley, New York, 1967

GUENTHER O, BUCHMANN A (1990)   Research issues in spatial databases. IEEE Bul-
letin on Data Engineering 13(4), pp 35-42

GÜNTHER O (1989)   Der Zellbaum. Ein Index für geometrische Datenbanken. In: Informatik Forschung und Entwicklung. Springer, Heft 4

GÜNTHER O, LAMBERTS J (1992)   Object-Oriented Techniques for the Management
of Geographic and Environmental Data, FAW Technical Report 92023, Sept.

GÜTING RH (1988) Geo-Relational Algebra: A Model and Query Language for Geometric Database Systems. In: J.W. Schmidt, S. Ceri, and M. Missikoff (eds), Advances in Database Technology - EDBT '88. Proceedings of the Intl. Conf. on Extending Database Technology, Venice, March , pp 506-527

GÜTING RH (1988a) Modelling Non-Standard Database Systems by Many-Sorted Algebras, Forschungsbericht Nr. 255, Fachbereich Informatik, Universität Dortmund, März 1988

GÜTING RH (1989) Gral: An Extensible Relational Database System for Geometric Applications. In: Proceedings of the 15th International Conference on Very Large Data Bases, Amsterdam, August 22 - 25

GÜTING RH (1991) Extending a Spatial Database System by Graphs and Object Class Hierarchies. Informatik Berichte Nr. 104, 1/91 FernUniversität Hagen

GÜTING RH, SCHNEIDER M (1993) Realm-Based Spatial Data Types: The ROSE Algebra, Informatik Berichte Nr. 141, 3/1993, FernUniversität Hagen

GUNTERMANN M (1994) Eine Rasterdarstellung für dreidimensionale Objekte in objektorientierten Datenbanken, M.Sc. thesis, Institut für Informatik III, Universität Bonn

GUTTMAN A (1984) R-Trees: A Dynamic Index Structure for Spatial Searching. In: Proceedings of the Annual Meeting ACM SIGMOD, Boston (MA), pp 47-57

HACK R, SIDES E (1994) Three-dimensional GIS: recent developments, ITC Journal, 1994-1, Delft

HAAS LM, CODY WF (1991) Exploiting Extensible DBMS in Integrated Geographic Informations Systems. In: O Guenther and H-J Schek (eds): Advances in Spatial Databases. Proceedings of the 2nd Symposium SSD'91, Zürich, Switzerland, LNCS No. 525, Springer, Berlin Heidelberg New York, pp 423-450

HÄRDER T, MEYER-WEGENER K, MITSCHANG B, SIKELER A (1987) PRIMA - a DBMS Prototype Supporting Engineering Applications. In: Proceedings of the 14th International Conference on Very Large Data Bases, Aug. 29 - Sept. 1, Los Angeles

HEALEY G, WAUGH TH (1994) Advances in GIS Research, Proceedings of the 6th International Symposium on Spatial Data Handling, 5th-9th Sept, Edinburgh

HERRING JR (1987) TIGRIS: Topologically Integrated Geographic Information System. In: Auto-Carto 8, pp 282-291

HERRING JR, LARSEN RC, SHIVAKUMAR J (1988) Extensions to the SQL query language to support spatial analysis in a topological database. Proc. GIS/LIS '89, San Antonio, TX, USA November

HOFFMANN CM (1989)   Geometric and Solid Modeling. Morgan Kaufmann Publishers, Inc., Santa Mateo, California

HOOP S DE, OOSTEROM P VAN (1992)   Storage and Manipulation of Topology in Postgres. In: Proceedings of the fourth European conference on Geographical Information Systems (EGIS)

JAESCHKE G, SCHEK H-J (1981)   Remarks on the Algebra of Non-First-Normal-Form Relations. In: Proceedings of the 1st. ACM SIGACT/SIGMOD Symp. on Principles of Database Systems, Los Angeles, Ca.

IFFLAND A (1994)   Implentation of Constructions with Space Primitives in Object-Oriented Databases, M.Sc. thesis, Institut für Informatik III, Universität Bonn

INGRAM KJ, PHILLIPS WW (1988)   Geographic information processing using a SQL-based query language. In: Proceedings of the 3rd Symposium on Spatial Data Handling, Sydney, Australia, Aug., pp 326-335

KAINZ W (1991)   Spatial Relationships - Topology versus Order. In: Proceeding of the 4th International Symposium on Spatial Data Handling, Zürich, pp 814-819

KELK B (1992)   3-D modelling with geoscientific information systems: the problem. In: A K Turner (ed), Three-Dimensional Modeling with Geoscientific Information Systems, NATO ASI 354, 29-38, Kluwer Academic Publishers, Dordrecht

KLESPER C (1994)   Die rechnergestützte Modellierung eines 3D-Flächenverbandes der Erftscholle (Niederrheinische Bucht). PhD Thesis at the Geological Institute, Universität Bonn. Published in: Berliner Geowissenschaftliche Abhandlungen, Reihe B, Band 22, 51 Abb., Berlin, S 117

KNIESEL G, ROHEN M, CREMERS AB (1991)   A Management System for Distributed Knowledge Base Applications. In: W Brauer, D Hernandez (eds), Verteilte Künstliche Intelligenz und kooperatives Arbeiten, Springer, pp 65 - 76

KOLLARITS S (1990)   SPANS 5.0, Benutzeranleitung FMM - TYDAC, Salzburg

KRAAK M-J, VERBREE F (1992)   Tetrahedrons and Animated Maps in 2D and 3D Space. In: BRESNAHAN et al. (eds), pp 63-71

KREMERS H (1991)   Umweltkartographie - Ein Teilgebiet der Umweltinformatik. Berliner geowiss. Abh. (C), 12, 71-76, Berlin, pp 71-76

LAURINI T, THOMPSON D (1992)   Fundamentals of Spatial Information Systems, Academic Press

LEISTER W (1991)   Modellieren mit Tetraederstrukturen, internal report No. 12/91, Aug. 91, Fakultät für Informatik, Universität Karlsruhe

LINNEMANN V, KÜSPERT K, DADAM P, PISTOR P, ERBE R, KEMPER A, SÜD-KAMP N, WALCH G, WALLRATH M (1988) Design and Implementation of an Extensible Database Management System Supporting User Defined Data Types and Funcions. In: Proceedings of the 14th International Conference on Very Large Data Bases, Los Angeles, August 29 - September 1

MAGUIRE DJ, GOODCHILD MF, D.W. RHIND (1991) Geographical Information Systems, Principles and Applications, Longman Scientific & Technical, Essex

MALLET JL (1992) GOCAD: a computer aided design program for geological applications. In: A K Turner (ed), Three-Dimensional Modeling with Geoscientific Information Systems, NATO ASI 354, Kluwer Academic Publishers, Dordrecht, pp 123-142

MÄNTYLA M (1988) An Introduction to Solid Modeling, Computer Science Press

MARX R (1994) A Topologically Based Data Structure for a Computer-Readable Map and Geographic System. Revista Cartografica, Nr. 46, Mexico (edited 1987), pp 101-112

MATSUYAMA T, HAO L V, NAGAO M (1984) A File Organization for Geographic Information Systems Based on Spatial Proximity. Computer Vision, Graphics and Image Processing, 26, pp 303-318

MEIER A (1986) Methoden der grafischen und geometrischen Datenverarbeitung.Teubner Stuttgart

MEIER A, LOACKER H (1987) POLY-Computergeometrie für Informatiker und Ingenieure. Hamburg: McGraw-Hill Book Company GmbH

MEISSNER U, WÖRNER J-D (1992, Hrsg) CAD und Expertensysteme im Bauwesen, Darmstädter Massivbau-Seminar

MOISE E (1977) Geometric Topology in Dimension 2 and 3, Springer, New York

MOREHOUSE S (1985) ARC/INFO a geo-relational model for spatial information. In Proceedings Auto Carto 7, Washington D.C., pp 338-357

MUKERJEE A (1989) A representation for modeling functional knowledge in geometric structures. Proceedings Knowledge Based Computer Systems, Bombay, Lecture Notes in Artificial Intelligence, Springer, Berlin Heidelberg New York, pp 192-202

NEUMANN K (1987) Eine geowissenschaftliche Datenbanksprache mit benutzerdefinierbaren geometrischen Datentypen, PhD Thesis, TU Braunschweig

NEWMAN W M, SPROULL RF (1981) Principles of Interactive Computer Graphics, McGraw-Hill Japan - Ltd.

NIEVERGELT J, HINTERBERGER H (1984) The GRID FILE: An Adaptable, Symmetric Multi-Key File Structure. ACM Transactions on Database Systems, 9(1), pp 38-71

NOACK R (1993)   Processing of Threedimensional Geometric Objects within the Extensible Object-Managment System OMS, M.Sc. thesis, Institut für Informatik III, Universität Bonn

ONTOS (1992)   ONTOS DB 2.2, Reference Manual, ONTOS Inc., Feb. 1992

OLIVER MA, WEBSTER R (1990)   A method of interpolation for geographical information systems. In: Int. J. Geographical Information Systems, Vol. 4, No. 3, pp 313-332

PAUL H-B, SCHEK H-J, SCHOLL MH, WEIKUM G, DEPPISCH U (1987)   Architecture and Impementation of the Darmstadt Database Kernel System. In: Proc. ACM SIGMOD, San Francisco

PAVLIDIS MG (1982)   Database Management for Geographic Information Systems. Proc. Nat. Conf. on Energy Resource Management, Vol. 1, pp 255-260

PERKHOFF A (1991)   Eine funktionale Transformationssprache zur Integration heterogener Datenbestände. In: Datenbanken in Büro, Technik und Wissenschaft BTW'91, Kaiserslautern

PEUCKER T, CHRISMAN N (1975)   Cartographic data structures. In: The American Cartographer, vol. 2, pp 55-69

PEUQUET DJ (1984)   A Conceptual Framework and Comparison of Spatial Data Models. In: Cartographica, Vol. 21, No. 4, pp 66-113

PFLUG R, KLEIN H, RAMSHORN CH, GENTER M, STÄRK A (1992)   3D Visualization of Geologic Structures and Processes. In: Lecture Notes in Earth Sciences 41 Berlin, Springer, 1992, pp 29-39

PIGOT S (1991)   Topological Models for 3D Spatial Information Systems. In: Proceedings of Auto-Carto 10, Baltimore

PIGOT S (1992)   A Topological Model for a 3D Spatial Information System. In: BRESNAHAN et al. (eds), pp 344 - 360

PIGOT S (1994)   Generalized Singular 3-Cell Complexes. In: Th Waugh and R Healeay, Proceedings of the 6th International Symposium on Spatial Data Handling, Edinburgh, pp 89-111

PILOUK M TEMPFLI K MOLENAAR M (1994)   A Tetrahedron based 3D Vector Data Model for Geoinformation. In: Proceedings of the International GIS Workshop "Advanced Geographic Data Modelling", AGDM'94, Delft

PIWOWAR JM, LEDREW EF (1990)   Integrating Spatial Data: A User's Perspective, Photogrammetric Engineering and Remote Sensing, Vol. 56, No. 11, Nov. 1990., pp 1497 - 1502

PREPARATA FP, SHAMOS MI (1985) (eds) Computational Geometry - An Introduction, Springer, Heidelberg et al.

PULLAR D (1988) Data Definition and Operators on a Spatial Data Model. ACSM-SSPRS Annual Convention, St. Louis, pp 197-202

QUERENBURG V B (1979) (Hrsg) Mengentheoretische Topologie, 2nd Edition, Springer, Berlin Heidelberg New York

RAPER J (ed) (1989) (ed) Three dimensional applications in Geographical Information Systems, Taylor & Francis, London 1989

RAPER JF (1992) Appropriate Design Criteria for a 3-D Geoscientific Mapping and Modelling System. In: Geolog. Jb., A122, Hannover, pp 109-119

RAPER J, RHIND D (1990) UGIX(A): The Design of a Spatial Language Interface for a Topological Vector GIS. In: Proc. of the 4th International Symposium on Spatial Data Handling, Zürich 1990, Vol. 1, pp 405-412

REQUICHA AAG (1977) Mathematical models of rigid solids. Techn. memo 28, Univ. Rochester, Rochester, New York, USA, Nov. 1977

REQUICHA AAG (1980) Representations for Rigid Solids: Theory, Methods and Systems. In: Computing Surveys, Vol. 12, No. 4, December, pp 437-464

REQUICHA AAG, VOELCKER HB (1982) Solid Modeling: A Historical Summary and Contemporary Assessment. In: IEEE Comp. Graph. Appl., 2,2, Los Ala mitos, CA, pp 9-24

REQUICHA AAG, VOELCKER HB (1983) Solid Modeling: Current Status and Research Directions. In: IEEE Comp. Graph. Appl., 3,7, Los Alamitos, CA, pp 25-37

RHIND DW (1992) Spatial data handling in the geosciences. In: A K Turner (ed), Three-Dimensional Modeling with Geoscientific Information Systems, NATO ASI 354, 13-28, Kluwer Academic Publishers, Dordrecht

RHIND D W, GREEN NPA (1988) Design of a geographical information system for a heterogeneous scientific community. In: Int. J. of Geographical Information Systems, Vol.2, No.2

RHIND D, OPENSHAWS, GREEN N (1988) The Analysis of Geographical Data: Data rich, Technology adequate, Theory poor, In: IV SSDBM

RUMBAUGH J, BLAHA M, PREMERLANI W, EDDY F, LORENSEN W (1991) Object-Oriented Modeling and Design, Prentice Hall, New Jersey

SALGE F, SMITH N, AHONEN P (1992)   Towards harmonized geographical data for Europe; MEGRIN and the needs for research. In: BRESNAHAN et al. (eds),Vol. 1, pp 294-302

SAKAMOTO M (1994)   Mathematical Formulations of Geological Mapping Process - Algorithms for an Automatic System. In: Journal of Geosciences, Osaka City University, Vol. 37, Art. 9, 21 fig., 2 tables, March, pp 243-292

SAMET H (1990a) (ed) The design and analysis of spatial data structures, Addison-Wesley, Reading

SAMET H (1990b) (ed) Applications of spatial data structures, Addison-Wesley, Reading

SCHALLER J (1988)   Das Geografische Informationssystem "ARC/INFO" und die mögliche Anwendung auf Geo-Daten,   ESRI, Gesellschaft für Systemforschung und Umweltplanung mbH

SCHEK H-J, WATERFELD W (1986) A Database Kernel System for Geoscientific Applications. In: Proceedings of the 2nd Symposium on Spatial Data Handling, Seattle

SCHEK H-J, WOLF A (1992)   Cooperation between Autonomous Operation Services and Object Database Systems in a Heterogeneous Environment. In: D Hsiao, E J Neuhold, R Sacks-Davis (eds), Proc. IFIP TC2/WG2.6 Conf. on Semantics of Interoperable Database Systems, DS-5, Lorne, Victoria, Australia

SCHEK H-J, WOLF A (1993)   From Exentsible Databases to Interoperability between Multiple Databases and GIS Applications, SSD'93, Signapore, LNCS 692, Springer, Berlin Heidelberg New York, pp 207-238

SCHNEIDER R (1993)   Geo-Datenbanksysteme - Eine Speicher- und Zugriffsarchitektur, Bibliographisches Institut Mannheim, Wissenschaftsverlag

SCHOENENBORN I (1993)   Management and Processing of Topological Structures in an Extensible Object Management System, M.Sc. thesis, Institut für Informatik III, Universität Bonn

SEEGER B, KRIEGEL H-P (1990) Design, Implementation and Performance Comparison of the Buddy-Tree. In: Proceedings DEXA'90, pp 203-207

SELLIS T, ROUSSOPOULOS N, FALOUTSOS C (1987) The R+-Tree: A Dynamic Index for Multi-Dimensional Objects. In: Proceedings of the 13th VLDB Conference, Grightion, pp 507-518

SHEPHERD IDH (1991) Information Integration and GIS. In: MAGUIRE et al. (1991), pp 337-360

SIEHL A (1988) Construction of Geological Maps based on Digital Spatial Models. Geol. Jb., A104, 2 Abb., Hannover, pp 253-261

SIEHL A (1993)   Interaktive geometrische Modellierung geologischer Flächen und Körper. In: Die Geowissenschaften, Bd. 11:10-11, 10 Abb., Berlin, pp 343-346

SIEHL A, RÜBER O, VALDIVIA-MANCHEGO M, KLAFF J (1992)   Geological Maps Derived from Interactive Spatial Modeling. In: From Geoscientific Map Series to Geo-Information Systems, Geologisches Jahrbuch, A (122), Hannover, pp 273-290

SILBERSCHATZ A, STONEBRAKER M, ULLMANN JD (1990)   Database Systems: Achievements and Opportunities, SIGMOD RECORD Vol. 19, No. 4, Dez., pp 6-22

SINHA AK, WAUGH TH C (1988)   Aspects of the implementation of the GEOVIEW design. In: Int. J. Geographical Information Systems, Vol. 2, No. 2, pp 91-99

SMALLWORLD GIS (1993)   Produktbeschreibung, SMALLWORLD Systems GmbH

SONNE B (1988)   Raumbezogene Datenbanken für kartographische Anwendungen. Geo-Informationssysteme, 1(1), pp 25-29

STIENECKE J (1995)   Räumliche Operationen und Versionen für ein Informationssystem zur Verwaltung geologisch definierter Geometrien, M.Sc. thesis, Institut für Informatik III, Feb. 1995, Universität Bonn

STONEBRAKER, KEMNITZ (1991)   The POSTGRES Next-Generation Database Management System. In: Communications of the ACM, 34:10, October

SVENSSON P, ZHEXUE H (1991)   Geo-SAL: A Query LAnguage for Spatial Data Analysis. In: O Guenther and H-J Schek (eds), Proceedings SSD'91, LNCS 525, 119-142, Springer, Berlin Heidelberg New York

SYSTEM9 (1992)   Systembeschreibung von SYSTEM9, COMPUTERVISION, BR9158, 4/92

TAMMINEN M (1982)   Efficient Spatial Access to a Data Base. In: ACM-SIGMOD, Orlando (FL)

TAMMINEN M, KARONEN O, MÄNTYLÄ M (1984)   Ray-casting and block model conversion using a spatial index. In: Computer-Aided Design, 16,4, Guildford, Surrey UK, pp 203-208,

TAMMINEN M, SAMET H (1984)   Efficient octree conversion by connectivity labeling. In: ACM Computer Graphics, 18(3), pp 43-51

TILOVE RB (1980)   Set membership classification: A unified approach to geometric intersection problems. In: IEEE Transactions on Computers c-29, pp 874-883

TOMLINSON RF (ed) (1972)   Geographical Data Handling. IGU Commission on Geographical Data Sensing and Processing, Ottawa

TURAN G (1982) On the Succinct Representation of Graphs. In: Discrete Applied Mathematics 8 (1982), pp 289 - 294

TURNER AK (1992) Three-Dimensional Modeling with Geoscientific Information Systems, NATO ASI 354, Kluwer Academic Publishers, Dordrecht

U.S.CENSUS (1970) US Dept. of Commerce , Bureau of the Census, The DIME geocoding system. In: Report No. 4, Census Use Study

VAN OOSTEROM P, VIJLBRIE T (1991) Building a GIS on top of the open DBMS "Postgres". In: Proceedings EGIS'91, pp 775-787

VAN OOSTEROM P, VERTEGAAL W, VAN HEKKEN M, VIJLBRIEF T (1994) Integrated 3D Modelling with a GIS. In: Proceedings of the International GIS Workshop "Advanced Geographic Data Modelling", AGDM'94, Delft

VIJLBRIEF T, OOSTEROM P (1992) The Geo^{++} System: an Extensible GIS. In: BRESNAHAN et al. (eds),Vol. 1, pp 44-50

VINKEN R (1988) Digital Geoscientific Maps -A Research Project of the Deutsche Forschungsgemeinschaft (German Research Foundation). In: R Vinken (ed), Construction and Display of Geoscientific Maps derived from Databases, Geol. Jb., Proceedings of the International Colloquium at Dinkelsbühl, FRG, Dec. 2-4, 1986, Geol. Jb., A104, Hannover, pp 7-20

VINKEN R (1992) From Digital Map Series in Geosciences to a Geo-Information System. In: R Vinken (ed), From Geoscientific Map Series to Geo-Information Systems, Proceedings of the International Colloquium at Wuerzburg, Sept. 9-11, 1989, Geol. Jb., A122, Hannover, pp 7-26

VOISARD A (1991) Towards a Toolbox for Geographic User Interfaces, In: Proceedings SSD'91, LNCS No. 525, Springer, Berlin Heidelberg New York, pp 75-98

VOISARD A (1992) Geographic Data Bases: From Data Models to User Interfaces. In French, PhD thesis, Universität Paris-Sud (Centre d'Orsay)

VOISARD A, SCHWEPPE H (1994) A Multilayer Approach to The Open GIS Problem. In: Proceedings of the 2nd ACM workshop on Advances in Geographic Information Systems, Dec. 1-1, Gaithersburg, Maryland

WATERFELD W (1991) Eine erweiterbare Speicher- und Zugriffskomponente für geowissenschaftliche Datenbanksysteme, PhD Thesis, Darmstädter Dissertation D17, TH Darmstadt

WATERFELD W, BREUNIG M (1990) Kopplung eines Kartenkonstruktionssystems mit einem Geo-Datenbankkern. In: W Pillmann, A Jaeschke (Hrsg), Informatik für den Umweltschutz, Informatik Fachberichte Nr. 256, Springer, Berlin Heidelberg New York, pp 344-354

WATERFELD W, BREUNIG M (1992) Experiences with the DASDBS Geokernel: Extensibility and Applications. In: From Geoscientific Map Series to Geo-InformationSystems, Geolog. Jahrbuch, A (122), Hannover, pp 77-90

WEICK W (1988) Beschreibung der CAD*I-Schnittstelle zum Austausch von Volumenmodellen In: H R Weber (ed), CAD-Datenaustausch und -Datenverwaltung, Springer, Berlin Heidelberg New York

WIDMAYER P (1991) Datenstrukturen für Geodatenbanken. In: G Vossen and K-U Witt, (Hrsg), Entwicklungstendenzen bei Datenbank-Systemen, Oldenbourg, München, pp 317-362

WOLF A (1989) The DASDBS GEO-Kernel, Concepts, Experiences, and the Second Step. In: A. Buchmann and O. Guenther (eds), Proceedings SSD'89, LNCS 409, Springer, Berlin Heidelberg New York, pp 67-88

WOLF A, DE LORENZI M, OHLER T, HAI NGUYEN V (1994) COSIMA: A Network Based Architecture for GIS. In: J Nievergelt, Th Roos, H-J Schek, P Widmayer (eds), IGIS'94: Geographic Information Systems, Proceedings of the International Workshop on Advanced Research in Geographic Information Systems, Monte Verita, Ascona, Schweiz, Feb. 28 - March 4, LNCS 884, Springer, Berlin Heidelberg New York, pp 192-201

WONHDBS (1990) Report on the Workshop on Heterogeneous Database Systems held at the Northwestern University, Evanston, Illinois. In: SIGMOD RECORD, Vol.19, No.4

WORBOYS MF (1992) A Model for Spatio-Temporal Information. In: BRESNAHAN et al (1992) Vol. 1, pp 602-611

WORBOYS MF (1994) Innovations in GIS I, Taylor & Francis

WORBOYS M, BOFAKOS P (1993) A Canonical Model for a Class of Areal Spatial Objects, SSD'93, Signapore, LNCS 692, Springer, Berlin Heidelberg New York

WORBOYS M, DEEN SM (1991) Semantic Heterogenity in Distributed Geographic Databases, SIGMOD RECORD, Vol. 20, No. 4, December, pp 30-34

ZALIK B, GUID N, VESEL A (1992) Representing Geometric Objects Using Constraint Description Graphs, IEA-92, Paderborn

ZHOU Q, GARNER BJ (1991) On the Integration of GIS and Remotely Sensed Data: Towards An Integrated System to Handle The Large Volume of Spatial Data. In: O Guenther and H-J Schek, Proceedings of the 2nd Symposium SSD'91, LNCS No. 525, Springer, Berlin Heidelberg New York, pp 63-74

Index

.

Lecture Notes in Earth Sciences